ENERGY IN AMERICA

ENERGY IN AMERICA

A Tour of Our Fossil Fuel Culture and Beyond

by

Ingrid Kelley

UNIVERSITY OF VERMONT PRESS

Burlington, Vermont

PUBLISHED BY UNIVERSITY PRESS OF NEW ENGLAND

Hanover and London

University of Vermont Press

Published by University Press of New England,

One Court Street, Lebanon, NH 03766

www.upne.com

© 2008 by University of Vermont Press

Printed in the United States of America

5   4   3   2   1

Library of Congress Cataloging-in-Publication Data
Kelley, Ingrid N.
Energy in America : a tour of our fossil fuel culture and beyond / Ingrid N. Kelley.
p. cm.
Includes index.
ISBN 978-1-58465-640-1 (pbk. : alk. paper)
1. Fossil fuels—United States.   2. Renewable energy sources—United States.
3. Sustainable living—United States.   4. Energy conservation—United States.   I. Title.
TP317.U5K45   2008
333.79—dc22   2008037290

University Press of New England is a member of the Green Press Initiative. The paper used in this book meets their minimum requirement for recycled paper.

This book is dedicated to everyone who knows that working on the solution is more than sufficient reason to get out of bed in the morning.

CONTENTS

ACKNOWLEDGMENTS

No book is created in a vacuum even though it occasionally feels that way. At those times, my sisters, Kristine Seaman and Linnea Nelson, were there to talk me through the highs and lows, to read initial drafts of my manuscript, and to offer their honest opinions. I am also very grateful to my parents, Helen and Arnold Nelson, for contributing their enthusiasm and feedback to the project and for still believing in me after all these years.

Friends and colleagues have offered their invaluable support as well. My thanks to Lynne Behnfield, Barbara Samuel, Taudy Smith, Anindita Mitra, Ralph Wilmer, Lee DeBaillie, and Steve Brick for reading early outlines or segments and responding with honesty and to Nancy Frank for offering detailed and valuable comments on the final manuscript. Special thanks as well to colleagues Kathryn Schiedermeyer and Joe Kramer for listening to me talk things through and to longtime friend John Porter for sharing his excellent sense of humor. I greatly appreciate Robin Seaman of Pirigraphics for her expert assistance with charts and illustrations and Cherie Williams for her artistry with the logo and the portrait. I am indebted as well to my wise friends Ana Larramendi and Kamerin Macmillion for so much insight and love, and to my permaculture teaching partner and co-conspirator, Peter Cooke, for his unique perspective and ongoing friendship.

Finally, my heartfelt thanks to Phyllis Deutsch for persuading me to take this on and for encouraging me to finish it and to her fine colleagues at the University Press of New England for being so helpful to a first-time author.

ENERGY IN AMERICA

Introduction

Energy is in the news as never before, and the issues cover a lot of ground. There is the growing U.S. dependence on foreign oil and the scramble to counter this with ethanol and other biofuels. Shrinking availability of natural gas and oil has forced up the cost of heating our homes. Global warming looms large as we contemplate rising sea levels and stranded polar bears, along with dramatic and destructive new weather patterns. International tensions are mounting over carbon emissions of rapidly developing nations and control of ever-tightening oil supplies. America is feeling the pressure of being the world's biggest consumer, and we feel a long way from living in a sustainable way.

Solutions are drawing headlines as well. Newspapers and magazines publish "green" issues offering carbon reduction ideas like using compact fluorescent lights, ENERGY STAR–qualified appliances,[1] and biking to work. We read about LEED® Silver, Gold, and Platinum commercial buildings that sip energy, and wind turbines in corn fields that offer farmers a new crop to sell. We hear about cities and towns across the nation declaring themselves the "greenest" or signing the U.S. Mayor's Climate Protection Agreement promising to support the goals of the Kyoto Protocol by reducing carbon emissions at the local level. The media are also fond of the "wow" factor, cheerfully reporting the most amazing new technologies. We hear about how a few square miles of solar panels built in the Nevada desert could power the entire country or how the hog manure in Iowa will be generating our power. Or we may hear a sound bite on TV about a scientist or engineer working on harnessing the power of tides, making hydrogen from sugar syrup, or using the energy of hundreds of feet dancing on a disco floor. These solutions are all great stories, but it's hard to believe any of them will make much difference against forces like Hurricane Katrina.

We know we need to begin serious reduction of carbon emissions and that these emissions come from using fossil fuels. We are realizing that the fossil fuel resources that power our culture are becoming simultaneously less easily available and more dangerous to live with. If we believe what the vast majority of scientists are telling us about global warming, we want to take action. We want to do something that will make a difference, but the anecdotal nature of the information we receive about the connection between global warming and energy is frustrating because we feel we don't know enough to take meaningful action. Furthermore, whether the excitement is about capturing and storing carbon emissions from coal-fired power, developing hydrogen cars, or generating electricity with landfill methane, our characteristically American obsession with technology makes it seem that only engineers and scientists, and possibly energy policy analysts, have the knowledge and skills to move us toward a less carbon-intensive energy infrastructure. However, most of what has shaped American energy infrastructure in the past has had little to do with anything resembling rocket science, and the future is unlikely to be any different. We will still be mixing up the politics and economics of public and private agendas with a dash of technical ingenuity and a pinch of audacious entrepreneurship to keep it interesting. The difference this time is that the stakes are higher. This time we're gambling with the climate that keeps us alive, and we all need to be conversant with the issues in order to participate in creating a sustainable energy future.

Our Fossil Fuel Culture

Our nation was born at the dawn of the fossil fuel age and literally grew up by flexing its entrepreneurial muscles in the development of fossil fuel technologies. Whether in wartime or peacetime, America's central position on the international stage could be directly attributed to the powerful infrastructure we have built with our vast coal, oil, and natural gas resources. Our fossil fuel culture has been steadily built one railroad track at a time, one power station, one steel mill or highway, over the last two hundred years, by Republicans and Democrats, public and private sector investment, and the collective will of the American people. There's no arguing with success.

No arguing, that is, until the phenomenon of global warming became impossible to deny. Suddenly, combustion of fossil fuels for transportation, power generation, industrial processes, and heating our homes has been abruptly transformed from a great solution to a huge problem. While this is true for all nations who use fossil fuels, it is particularly tough for the United States, because we produce about 22 percent of the world's carbon dioxide (CO_2) emissions, which calculates out to be the highest amount per person in the world. We have based our cities, our businesses, and our lifestyle on the convenience and relatively low cost of gasoline, coal-fired electricity, and plastic. It's no wonder that the specter of global warming appears so devastating: it threatens the roots of our culture.

The challenge is tremendous, not to say overwhelming. We can no longer leave the decisions about our energy future to the technologists and energy industrialists. City officials, architects, planners, urban designers, engineers, and community activists are feeling the pressure to address the complexities of that simple-sounding charge: "reduce your carbon footprint." There is currently a mad search for effective strategies to accomplish this feat, for the magic list of "best practices" that will reduce carbon emissions, not only quickly but economically. We want to change the system, but how can we change what we have if we don't understand what that is? How do we know the difference between what we should keep and what no longer serves us?

This is the reason I wrote *Energy in America*. Friends and colleagues assumed I would write a how-to book or a collection of case studies of successful clean energy projects so people would have some good ideas about what to do. Such a book would be extremely useful in our effort to make changes, and someone else is undoubtedly writing it. The book I chose to write, however, is one that gives people some background about how the world of energy operates in America so they could pick up that book of great examples and evaluate them intelligently within the context of everyday reality.

I have worked as an energy professional for over twenty years, starting as a mechanical designer on heating and cooling systems, and gradually broadening my interests to include energy efficiency, renewable energy, green building, permaculture, and sustainable community planning. Then I moved from design into policy and public outreach, writing a lot of fact sheets and articles, giving many presentations, and serving on a variety of energy policy and sustainability task forces. I have worked with utility professionals, state government energy program managers, engineers and architects, renewable energy installers, alternative construction experts, permaculture designers, and environmental scientists. In 2001, I earned my Masters in Community and Regional Planning, hoping to tie all this experience together to become a sustainable community planner. Then I discovered something disconcerting. When I entered the world of sustainable community design, I found that this group of professionals was not really talking about energy, and that many of them, even those very knowledgeable about closely related sustainability and environmental issues, know very little about the energy infrastructure we have come to rely upon. Yet it is our sources and uses of energy that we will need to reexamine and change if we are to address global warming effectively.

With *Energy in America,* I am sharing the knowledge and insights that have come my way over the last twenty years as I have sought to understand the close connections between energy and sustainability. The book is organized loosely by sector and chronology, starting with what we have developed in the fossil fuel industries and the electric utility grid, and moving forward to more recent concerns with energy efficiency and renewable energy. I conclude with chapters on local energy planning and national and regional global warming policy. Underlying these factual narratives, however, are three general ideas I feel are even more important for anyone wanting to understand the energy picture in America today. First, engineering expertise is handy, but hardly the most important skill to have. Second, energy sources or technologies do not win general acceptance based on their performance alone, and last, considering how energy industries have developed in the past, it should be quite clear that nothing about the future is carved in stone or cast in plastic. We do have choices, which are limited only by our imaginations and our political will.

It's Not About the Technology

Energy can be a very technical subject, as any engineer would tell you, but when we talk about transforming our current, unsustainable energy infrastructure into one that will still support us in a livable climate seven generations down the line, the technological challenge is hardly the greatest. The most significant barriers to creating a clean energy future are political, and the primary impediments are the age-old, archetypal opponents of progress we know so well: short-sighted economic self-interest, lack of knowledge, and an unwillingness to change established ways, even in the face of alarming evidence. Overcoming these barriers is not easy, and it is not technological ingenuity that will make the difference. Leadership must come from public officials, design professionals, planners, and community activists who inform themselves about energy issues and infrastructure and recognize the power in such tools as legislation, regulation, and public education to open the economic and social potential of a clean energy future.

There Is No Such Thing as Technical Darwinism

We have the most convenient energy system, and one that is envied around the world. All we need to know about are the gas pump, the thermostat, and where the wall outlets are, and our lives move smoothly along. We no longer have to worry about chopping firewood for cooking or heating bath water, shoveling coal into a furnace to heat our homes, or making candles to provide light at night. It is therefore tempting to think there is some sort of technological "survival of the fittest" principle at play in the sense that energy sources and technologies have competed over time and those that work the best have prevailed, giving us the wonderful system we have today. Do we dare tinker with this paradigm, sanctioned as it is by the wisdom of time?

First, let us not confuse our sources of energy with the technologies that deliver the convenience of using them. Neither power plants nor vehicles require fossil fuels to do their jobs, and the electricity from a wind turbine is identical with that from a coal-fired generator. We don't need to give up convenience, even though it was primarily the product of our fossil fuel culture.

As for the question of fossil fuels versus renewable energy, our future energy scenario will not be dictated by a Darwinian "natural selection" of technical performance, with the selection made by Adam Smith's "golden hand"—market forces objectively determining what is most economical. Political will and economic reality will continue to drive our energy choices as they have always done in the past. The fossil energy industries have long been subject to price and supply manipulation both internally and externally through such strategies as government subsidies, regulated utility rates, and cartel agreements. The OPEC nations have often manipulated the oil market the way the Federal Reserve manages the prime interest rate, frequently in the interests of international economic stability. The federal government's reluctance to put a price tag on the environmental impacts of coal-fired power has kept utility rates artificially low and the relative cost of renewable energy high. If we look to nations using substantial renewable energy sources such as Denmark, Spain, or Germany, we find commitment to subsidize solar and wind power at the national government level to be not unlike our own government's subsidies of fossil fuels.

Technology follows economic opportunity, and the economics of energy do not exist in a vacuum. Outside the United States, renewable energy is seen as the major growth industry of the twenty-first century, while here in America we are seriously considering making transportation fuel from coal. There is nothing guaranteed about our energy future, and there are no laws of physics or economics that will dictate what choices we ultimately make.

It's All Right to Imagine a Very Different Energy Future—We Do Have Choices

A new age for energy is beginning, and though its presence is already felt, those of us alive today can only dimly imagine what it will mean for the future. Fossil fuels are what we know and how we get things done today. Our neighborhoods, our industry, our landscape, and our economy exist in their present forms because of fossil fuels. To us, fossil fuels signify progress and prosperity, but living this way is no longer sustainable. Changing our perceptions can help us move to something new. The new energy age can begin when we fully realize that we can choose something different. Choosing a path of sustainable design and planning for our communities will open up unforeseen opportunities for developing the energy infrastructure of the future.

Fortunately, we've learned a lot about energy technology during the last two hundred years. The press of industrial progress has produced a remarkable number of technological innovations, along with the advanced manufacturing processes required to perfect them. Using renewable energy sources will mean more sophisticated and efficient technologies than we had before. The internal combustion engine and the electric motor would likely not have been so well developed without the energy abundance of the fossil fuel age, and yet neither requires gasoline or coal to operate. Furthermore, there is no law of physics or otherwise that says we must generate electricity by combusting fossil fuel.

We have also learned a lot about providing ourselves with shelter that requires less energy to heat and cool. Insulation, efficient windows, and tight construction methods create homes and commercial buildings that use far less energy than they did even thirty years ago. With reduced energy needs, these structures can make more efficient use of green power and other renewable energy techniques such as passive solar design for heat and light, or solar thermal panels that heat water or air.

If fossil fuels allowed us to get global, their decline will mean that in many ways we will need to get local once again, something many see as an important criterion for designing a sustainable way of life. The technological skills we acquired during the fossil fuel age will allow us to create efficient and cosmopolitan communities powered by clean energy where we participate globally while living locally. Already, renewable energy industries are creating new local employment and investment opportunities, making ownership available to a broad cross-section of the community and creating the potential for a flexible and secure local energy infrastructure.

A clean or sustainable energy future cannot simply mean plugging new sources of renewable energy into the conventional energy infrastructure. We cannot keep the traditional centralized, hierarchical energy delivery model and expect to move easily into a sustainable future. Energy production and use must employ the latest renewable energy technologies, but it must interrelate with the social, political, economic, and environmental visions of sustainable communities as well. New models for energy development and distribution are needed, and we can decide to integrate these models into the philosophy and practice of sustainable community design.

Connecting Energy and Sustainability

Most people agree that definitions of sustainability have something to do with leaving the world no more depleted than we found it, or that we conduct our lives in such a way that we leave behind for our children's children no destructive sign of our passing. When we look to where sustainability starts, perhaps we point first to reviving a clean and healthy natural environment where there is room for a diversity of species to flourish, and where human impact is minimal and respectful. Or maybe our concern is first for homes, buildings, and other structures to be made of renewable and nontoxic materials and for transportation alternatives to make neighborhoods pedestrian-friendly, densely communal, and economically diverse. Being able to feed ourselves reliably and have clean drinking water are the issues of highest priority for some sustainable community designers. Or perhaps sustainability starts with environmental and social justice, strong local economic systems, and revitalization of community values.

Every one of the sustainable community concerns mentioned above is intimately entwined with our energy resources and how we use them. Energy is so big we don't see it, because energy is not part of the picture but the frame. If we wish to create sustainable communities, we must recognize that our relationship with energy is a major issue, and clean energy is more than simply one item on a list of green strategies. In order to make the kind of systemic changes that will be necessary, we need to understand the energy choices we made in the past, as well as the alternatives being explored right now and the policies and players in the market today.

The challenges of global warming and declining oil supplies have galvanized the energy industries in America. We are watching a giant struggle as the fossil fuel industries, the nuclear power industry, and renewable energy entrepreneurs jockey for political and economic support. There is no unilateral agreement that renewable energy development is the best investment strategy, and even among renewable energy developers there is competition for financial investment and public attention. The coal industry proposes to make liquid fuel from coal and is promoting its product as a potential "alternative" transportation fuel to lessen our dependence on foreign oil. And with "clean coal technology," the industry is promising ways we can continue to use coal for generating electricity while cutting carbon emissions through carbon storage technologies. The international demand for oil will soon put us into competition with rapidly developing nations like India and China as we seek new sources. Natural gas supplies are anything but secure either, and natural gas is much more expensive than oil to move from continent to continent.

America uses a lot of energy and is planning to use a lot more. No one

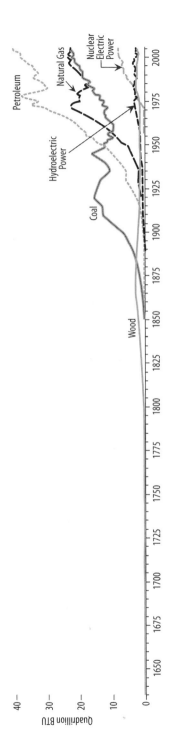

Fig 1. Energy Consumption by Source, 1635–2005. U.S. Department of Energy, Energy Information Administration, *Annual Energy Review 2006*, DOE/EIA-0384 (2006)

is certain exactly where it will come from, or whether we will have to sacrifice the quality of the environment to get it. The tremendous uncertainties of global warming and the tumult they have caused in our picture of ourselves and our communities have made us all involved in the outcome.

This book presents a view of our nation's energy picture from the perspective of one person who has been involved with energy and sustainability for a long time. Energy colleagues will have bones to pick about what I have left out and what I say about what is actually included. But that is all right, because I didn't write this book for them. Most energy professionals are specialists. They know about utility management, oil extraction, solar panel installation, biofuels manufacturing, energy efficiency, coal mining, or any of hundreds of other energy-related professions. Energy is a huge subject, and this book is intended merely as a starting point—the nickel tour, as it were. It is a personally acquired macroview of energy in the United States, which I hope will provide you with a broad enough perspective to start asking your own questions and finding the answers.

I decided not to include a lot of energy statistics, although there are a few that seemed to augment the narrative rather than dull the senses. My intention is to encourage exploration of the incredible information resources available on the web, and I have included a list of web resources at the end of each chapter. I rely on many of them frequently myself. I'm going on the assumption that, if you need the numbers, you will be able to find them easily enough. This is not a technical book either, although I will attempt to explain and describe some technologies and concepts. *Energy in America* is mostly about our relationship as a nation with energy—fossil fuels, electricity, energy efficiency, and renewable energy—and the way this relationship has developed and continues to shift, particularly in light of new information about global warming.

I admit to a strong bias toward renewable energy, local economies, and respecting the environment that supports our lives. I have also become convinced that the capital-intensive fossil and nuclear energy industries we live with now no longer serve the planet's best interests. I have found, however, that energy issues are not simple standoffs between those who wear black hats and those who wear green ones. Nor are the largest challenges we face about technology, or even about the energy resources themselves. Our beliefs, desires, fears, and willingness to change the parameters of our comfort zone will have far greater impact on our ability to move to clean sources of energy in the future.

Dealing with global warming is going to be complicated, messy, and expensive. The degree to which this will be true depends entirely on us. A sustainable world is one in which we will all have to take more personal responsibility to learn about the choices we have, the trade-offs we will be forced to make, and the tremendous opportunities there are to create cleaner, simpler, and more equitable ways of living. To meet the challenge of global warming and to create a sustainable society, the answer is simple. We need to become better people. It will be challenging to change attitudes and habits we've spent generations acquiring as a result of the fossil energy infrastructure. I believe that only by understanding the systems we've built can we transcend our fossil fuel culture and choose another path, because we can't tear something down and replace it until we understand why it was created and what it has accomplished.

NOTE

1. Products/Homes/Buildings that earn the ENERGY STAR prevent greenhouse gas emissions by meeting strict energy efficiency guidelines set by the U.S. Environmental Protection Agency and the U.S. Department of Energy. ENERGY STAR is a registered mark owned by the U.S. government.

RESOURCES

The following organizations provide information about a broad range of energy, environmental, and sustainability concerns.

Natural Resources Defense Council
An independent nonprofit environmental action organization that offers information about a variety of energy issues and impacts.
http://www.nrdc.org

Union of Concerned Scientists
An independent nonprofit, this organization is known as a reliable source of independent scientific analysis of environmental and energy issues.
http://www.ucsusa.org

Worldwatch Institute
Worldwatch Institute has been offering rigorously researched information about global environmental and social trends since 1975, in the interests of a sustainable and just international society.
http://www.worldwatch.org

Pew Center on Global Climate Change
(www.pewclimate.org)

CHAPTER ONE:

Fossil Fuels

We had a coal furnace in the house where I grew up. The coal bin was a dark and sooty room at the back of the basement, with a board-covered window in the wall next to the driveway. The coal truck would back in and empty coal down a chute through the window into the coal bin. My father would go down to stoke the furnace using a high-sided coal shovel. I still remember watching him do this, and I remember smelling the coal and the heat. Every once in a while, he would clean out the clinkers with the same big shovel and fill up a trash can.

Eventually my parents replaced the big, old coal furnace with a trim, new gas furnace. Suddenly there was room in the basement for my father to have his workshop, and he didn't have to do anything to the furnace anymore. I was fascinated with the large patch of new concrete on the floor that showed where the coal furnace had been. We cleaned out the coal bin, and that's where we put the new freezer. Suddenly we were modern. We didn't can fruits any more, we froze them. And it was much quieter in the basement.

—Childhood memories of the author

Coal, oil, and natural gas are woven into our lives and memories, so to understand our energy picture we must start with them. The United States uses a lot of energy and most of it comes from fossil fuels. The United States is also a longtime leader in developing fossil fuel resources and technologies. Today we rely on coal, oil, and natural gas for about 85 percent of our total energy needs, which includes over half of our electricity and almost all our transportation fuel. It is hard to imagine living without these energy sources. We could call our time the Fossil Fuel Age because it had its beginnings in the nineteenth century and we are seeing now that it will likely end in the twenty-first.

Fig. 2. Oil rigs on Venice Beach, 1945. The City of Venice, California, sat atop a small but rich oil deposit that was discovered in 1929. Over 450 oil rigs sprouted in backyards and along the scenic beach, today the location for its famous Ocean Front Walk. The last oil was pumped from Venice wells in the 1970s. Photo by Herman Schultheis, 1945. Reproduced with permission of the SE-CURITY PACIFIC COLLECTION / Los Angeles Public Library.

Coal, oil, and natural gas are called fossil fuels because they were formed from layers of carbon-based organic matter that had been plants and simple marine creatures many millions of years ago. Buried under sediment, these organic layers were concentrated by time, pressure, and terrain into various carbon deposits containing significant amounts of energy, which those ancient swamps and forests soaked up from the sun. When we began to extract and use these deposits, as coal, oil, or natural gas, human society was profoundly changed forever. For nations embracing fossil fuels, economic productivity increased exponentially, which permanently raised the bar for measurement of competitiveness among nations. Fossil fuel resources are so remarkable that they have brought out the best and the worst in people, from economic prosperity for millions and technological marvels like electricity, wonder drugs, and air travel, to war and political strife, greed, repression, and environ-

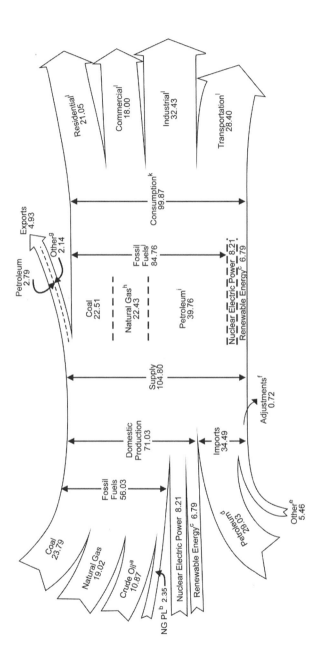

a Includes lease condensate.
b Natural gas plant liquids
c Conventional hydroelectric power, biomass, geothermal, solar/PV, and wind.
d Crude oil and petroleum products. Includes imports into the Strategic Petroleum Reserve.
e Natural gas, coal, coal coke, fuel ethanol, and electricity.
f Stock changes, losses, gains, miscellaneous blending components, and unaccounted-for supply.
g Coal, natural gas, coal coke, and electricity.
h Natural gas only; excludes supplemental gaseous fuels.
i Petroleum products, including natural gas plant liquids, and crude oil burned as fuel.
j Includes 0.06 quadrillion Btu of coal coke net imports.
k Includes 0.06 quadrillion Btu of electricity net imports.
l Primary consumption, electricity retail salers, and electrical system energy losses, which are allocated to the end-use sectors in proportion to each sector's share of the total electricity retail sales. See Note, "Electrical Systems Energy Losses," at end of Section 2.

Notes:—Data are preliminary. • Values are derived from source data prior to rounding for publication. • Totals may not equal sum of components due to independent rounding. Sources: Tables 1.1, 1.2, 1.3, 1.4, and 2.1a.

Fig 3. Energy Flow 2006 (Quadrillion BTU). U.S. Department of Energy, Energy Information Administration, *Annual Energy Review 2006,* DOE/EIA-0384 (2006)

mental disaster. The world supply of coal, oil, and natural gas has been so readily available and so adaptable to meeting our desires that, by now, we can hardly distinguish between what these fuels do and what they don't do in our lives.

Fossil Fuels Are Us

The United States continues to link its national identity to use of fossil fuels. It's true that the portable and flexible nature of fossil fuels has allowed development of a massive, centralized energy infrastructure that it would indeed be expensive to change if we wanted to move toward a more local, renewable energy scenario. However, as long as federal policy regards fossil fuels as vital to the well-being of the United States, the primary energy challenge will be neither economic nor technological, but political.

Our Fossil Fuel Age closely coincides with the history of the United States as a nation, so it should come as no surprise that fossil fuels have had particular influence on the formation of our culture. (Appendix B, Timeline: Seven Generations of Fossil Fuels, begins charting fossil fuel use in America starting before 1775.) The Industrial Revolution began in England where coal was mined and used to expand the manufacturing sector, making goods from the raw materials imported from that country's many colonies. Britain's colonial tradition drove expansion of fossil fuel use, but it retained much of its class structure. It was the fledgling United States that fully embraced the social and cultural potential created by these highly concentrated carbon fuels. Somehow the size of the continent and the dreams of a democratic society provided the appropriate scale for this new power source that could replace human and animal muscle a thousand times over and provide the means to light up the night.

Fossil fuels made possible the prosperity that equalized men of commerce and ambition with those of rank and royal blood. Coal, and later oil, could multiply the productivity of animal or human muscle many times over. Effort invested could yield more than a simple living, and ingenuity could bring riches. An explosion of fossil fuel technology supported a raw and competitive form of democracy that rewarded the inventive entrepreneur and dashed the pretensions of old world values. In the nineteenth and early twentieth centuries, America's aristocracy included coal barons, oil barons, and railroad tycoons. They were followed by automobile manufacturers, industrial, chemical, and fertilizer magnates, and fortunes built on selling the many resulting products. What coal and oil could do was larger than anything previously known—transportation, manufacturing, and the heating and lighting of cities.

In twenty-first-century America we have sports and entertainment superstars who would be almost completely unknown without the electronic media gadgetry currently brought to us in many ways by coal and natural gas. Fossil fuels have made most American fortunes possible, and they continue to do so today. Coal, oil, and natural gas have given us speed, convenience, and unceasing growth, not to mention a period of world dominance. It's been an incredible ride despite the pollution, the frequently less than humane working conditions for miners and factory workers, and the political power brokering that always results from too much money and power in the hands of too few. However, for almost two centuries, the downside of using fossil fuels couldn't compete with the upside for our attention.

Now we are asking ourselves, how deep is our dependence on the fuels that built America? The problem is that fossil fuels are not just fuels. Sure, we use a lot of gasoline, diesel, and aviation fuel to get around, a lot of coal to generate electricity and make steel and natural gas to heat our homes and dry our clothes. We are using increasing amounts of natural gas in many other ways as well because it is cleaner than the other two. But thousands of products we use everyday are made from coal, oil, or natural gas, and most of these products, many of which we now consider essential, didn't even exist a hundred years ago. Many products were previously made of plant or animal materials. For example salicin, a compound in willow bark, was a well-known herbal pain reliever. By the mid-nineteenth century chemists had distilled a similar chemical called acetylsalicylic acid from coal tar to make the first synthetic aspirin. Today aspirin is made from petroleum.

Global Warming and Climate Change

Before we take a closer look at the distinct history and character of coal, oil, and natural gas, let's quickly review the environmental downside they share: their release of greenhouse gasses when they are burned. These gases include carbon dioxide, methane, nitrogen oxide, and ozone. The buildup of greenhouse gasses in the atmosphere causes global warming.

Global warming occurs when heat from the sun is trapped in the earth's atmosphere by the presence (primarily) of carbon dioxide. While this phenomenon helps maintain a balanced temperature on the planet, the addition of more carbon dioxide to the mix means that more heat is retained and the overall temperature rises. Carbon dioxide is essential for life, but increasing the amount of it in the atmosphere disturbs the earth's ability to shed excess heat from the sun, which over time can profoundly alter the climate. Carbon dioxide is not the most potent of the gasses referred to as "greenhouse" gasses, but it makes up 83 percent of them and that's the problem. Most human-generated carbon dioxide emissions come from the combustion of fossil fuels. Many scientists are saying we need to cut carbon emissions by 80 percent before 2050 to keep rising global temperatures from dangerously destabilizing world climate patterns.

The reason fossil fuel combustion raises levels of carbon dioxide so significantly is that coal, oil, and natural gas are retaining all the carbon dioxide stored by the swampy plant matter that originally formed them, and when they are burned it is released. There is now scientific consensus that our release of such great quantities of CO_2 during the last two hundred years is likely to have contributed to changes in the natural temperature patterns human beings have experienced for the last several thousand years.

The phrases "global warming" and "climate change" get used interchangeably, but they are not the same concept. Scientists first started talking about global warming to describe the overall phenomenon of the earth's atmosphere gathering more heat when the sun's energy becomes less able to escape. However, they realized the phrase "global warming" makes it sound like a gradual and evenly distributed happenstance which does not express the potential complexities. The phrase "climate change" came into use to address the unpredictability of what may occur, such as more frequent and highly destructive weather events like powerful hurricanes or rainfall that produces heavy flooding. There may also be more gradual patterns that will make some areas drier or colder and others more hot or humid. One pattern already in evidence is the melting of glaciers and polar ice, which has altered wildlife habitat in colder climates and has begun to bring about rise in sea levels, causing great concern among low-lying island and coastal nations. A sobering possibility is that rising temperatures will reach a point where they trigger huge and devastating changes in a brief period of time.

The International Perspective on Climate Change

The coal, oil, and natural gas industries comprise a vast and complex international network that includes every nation in the world regardless of size or political agenda.

Therefore, climate change is clearly an international issue. Looking at how other governments have chosen to address it provides an interesting contrast to U.S. policy. Twenty percent of Denmark's electricity comes from wind power, and Germany is working toward producing 25 percent of its electricity with renewable energy, primarily solar, by 2025. Germany's proactive clean energy policies are all the more remarkable considering that they have long had the expertise to produce transportation fuel from domestic coal sources—they did this during World War II when the Allies cut off their supply of oil. Europe and Scandinavia in general have long had high gas prices, small cars, and quality public transportation. About one-third of Brazil's transportation fuel is biofuels, made primarily from sugar cane. India, China, and Spain have become world leaders in wind turbine and solar panel manufacture, and China's manufacturing potential could tip solar electricity into the main stream. To meet the tremendously challenging transition ahead of us, international consensus would clearly move us in the direction of developing renewable energy.

The Difference Between Peak Oil and Climate Change

Our dependence on foreign oil—our inability to produce all the oil we need—is colored by the reality that oil is not as easily available in other countries either as it once was. In 1956, a geoscientist named M. King Hubbert, who worked for Shell Oil, developed his theory that production of oil follows a bell curve, and that U.S. production would "peak" between the 1960s and 1970s, when we would reach the point where we are using the easily accessible oil more quickly than we are finding it. His hypothesis proved to be correct, and his calculations are now being applied to international supplies. The phenomenon of "peak oil" is explained in greater detail in the segment of the chapter covering oil. Peak oil is mentioned here because it has intertwined in the public mind to some extent with climate change. They are not at all the same issue even though reducing the use of fossil fuels would contribute significantly to solving both. The peak oil issue has sparked a lively national conversation about our dependence on the world oil supply and our necessary involvement with volatile political regimes to keep our highways humming. However, the objectives of those concerned about peak oil do not necessarily align with the priorities of those who would reduce the nation's carbon footprint. Proponents of "energy independence" focus on finding domestic sources of vehicle fuels regardless of their environmental impact, emphasizing the necessity to keep the economy moving. Looking only to find something local to replace petroleum for our gas tanks, such as manufacturing vehicle fuel from coal, ignores the issue of carbon emissions, which will only get worse if overlooked. Domestically produced transportation fuels could instead include ethanol and biodiesel made from corn and other biomass feedstocks, or we may rely on electric-powered transport. It is important that we keep an eye on the environmental impact of any proposed solutions to reducing petroleum use, whether the challenge is our dependence on foreign supplies or the issue of peak oil.

The Three Fossil Fuels

Although they share a common origin, each of the three primary fossil fuels has its own characteristics and issues. We have a different relationship with each one. We appear to have plenty of coal for generating electricity, but mitigating its impacts on our health and on the earth's atmosphere will be expensive. Oil is less environmentally destructive, but our transportation-intensive society is using enough of it to cause a similar level of damage to that of coal, and now we must figure out how to replace the diminishing supply with something as available but less destructive. Natural gas is pretty clean and very handy but that's the problem. So many different users are depending on it to supply them in the future that it's become like the income tax rebate check you spend several times over before it finally arrives. Natural gas cannot give us everything we need. There is only so much natural gas available and it is much more expensive than oil to move from one continent to another.

Coal: The King

Most Americans don't think much about coal because it seems sort of old-fashioned. Coal brings up images of grimy industrial cities and gritty railroad yards of days gone by. Back in the "gaslight" era, coal was the source of the gas people used for streetlights and in homes and buildings. It's true that we've replaced coal with oil for trains and ships, and we now use natural gas for heating and cooking. Even heavy industry has switched to natural gas for many processes, and a lot of factories rely mainly on electricity. In today's society, one could easily go through life without ever seeing coal, much less thinking about it. But coal is still vital to our way of life.

Coal is essentially a rock that burns, and it is the rock upon which the Industrial Revolution was built. Coal has been called "King Coal" for good reason. At the turn of the twentieth century we used more coal than anything else for transportation, electricity, industrial processes, and for heating homes and businesses. We still burn coal today to generate over half of our electricity.

The United States has some of the largest coal reserves in the world. Coal deposits in the Eastern states were formed primarily in saltwater marshes about 300 million years ago. Generally speaking, Eastern coal is harder, contains less moisture and more sulfur, and is farther underground than coal deposits in the Western states. Western coal is much younger, being about 140 million years old. It is softer and closer to the surface, and does not have the energy content of Eastern coal. However, its freshwater origins make it much lower in sulfur content.

There are four types or "ranks" of coal, which are classified by their carbon content and chemical makeup; the harder the coal, the higher the carbon content and the lower the level of moisture. The hardest coal is anthracite, which contains from 86 to 98 percent carbon. This is the cleanest burning type and has been traditionally used both for steel making and for space heating. It was anthracite coal that was shipped into the cities for use in homes because it produced comparatively little smoke. Anthracite coal deposits in the United States are located almost entirely in Pennsylvania (also the home of the American steel industry), but these deposits have been mostly exhausted. Bituminous coal is the second "hard" coal, and the most common type found in the United States. Its carbon content ranges between 45 and 86 percent. Bituminous

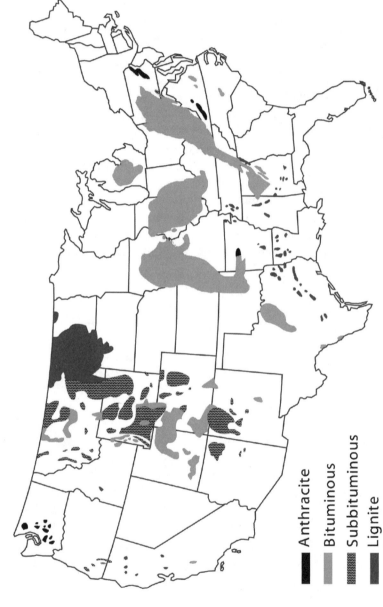

Anthracite

Bituminous

Subbituminous

Lignite

Fig 4. Coal Deposits in the United States, 2006. Based on information from the U.S. Energy Information Administration. Map courtesy of the National Mining Association

The Fossil Fuel Family		
Coal	**Oil**	**Natural Gas**
Coke	Kerosene	Compressed Natural
Coal Gas	Gasoline	Gas (CNG)
Kerosene	Propane	Liquified Natural Gas
	Heating Oil	(LNG)
	Aviation Fuel	Propane
	Diesel	Butane
	Petroleum Coke	

coal is found in the Appalachian Basin, which includes West Virginia, Kentucky, Pennsylvania, and Ohio. It also occurs in the Illinois Basin (Illinois, Indiana, and western Kentucky). Generally speaking, Appalachian coal is lower in sulfur content than Illinois Basin coal. Bituminous coal is used primarily for power generation, cement manufacturing, and making coke for smelting iron and making steel.

The two "soft" coals are sub-bituminous and lignite. Sub-bituminous coal yields 35 to 45 percent carbon content and lignite is the softest and lowest at 25 to 35 percent. The greatest source of sub-bituminous coal is currently the Powder River Basin in Wyoming, a 12,000-square-mile coal field in the northeastern part of the state. Lignite is mined primarily in Texas and North Dakota, but Montana also has extensive reserves. Even though these two types are significantly lower in energy content, they have become very popular for generating electricity because their sulfur content is much lower than that of Eastern coals, allowing power plants to meet federal pollution standards.

The Wyoming coal has an additional advantage. The Powder River Basin is flat, unpopulated prairie, and with the coal deposits being relatively close to the surface they are perfectly suited to economical, high-volume surface mining. The price of Wyoming coal is very competitive despite the longer transport distance. As recently as 1968, Western states were producing only about 5 percent of the nation's coal. In 1988 Wyoming became the highest producing coal state in the coun-

The Formation of Fossil Fuels

Coal, oil, and natural gas are organic in origin and are all products of ancient biological and geological activity. They were formed between 100 and 400 million years ago from decaying plant and animal matter. It could be said that the energy in fossil fuels came from the sun, was collected by ancient plants and animals, and was then transformed and stored as carbon, a ready source of fuel today.

Coal

Coal started as peat, a fuel still used today in some parts of the world. Many coal deposits were formed during the Carboniferous Period, 354 to 290 million years ago, a time characterized by vast areas of swamps filled with lush vegetation and aquatic plant matter. Peat bogs formed primarily in river valleys from the accumulations of vegetation and organic matter as it decayed. This organic matter was broken down by anaerobic bacteria, which thrive without oxygen. Over time these bogs were filled in and buried by sediment. Heat and pressure compressed the peat and removed the moisture, and the resulting chemical changes produced seams or layers of hydrocarbon deposits. The purity of the coal in these deposits depends on the intensity of the heat and pressure, and the duration of these conditions. Coal is made up of carbon, hydrogen, oxygen, nitrogen, and sulfur. High sulfur coal was formed originally from saltwater vegetation, and low sulfur coal comes from ancient, freshwater vegetation deposits.

Oil

The process is similar for oil, but the organic matter was primarily ocean plant life and one-celled sea creatures that died and collected on the ocean floor. These oceanic deposits were broken down by the same type of anaerobic bacteria found in peat bogs, transforming it to a thick, black substance called kerogen. Once again, sand and silt buried the layers of kerogen, and heat and pressure transformed it into oil. Surprisingly, the actual process is still something of a mystery, but theoretically, the kerogen undergoes further transformations through the

The Formation of Fossil Fuels *(continued)*

presence of heat, resulting in various molecular weights of oil and gas. These substances work their way upward, collecting in porous rock or becoming trapped and pressurized under layers of impermeable rock like shale or limestone. Depending on the geological conditions and history, some deposits will contain just oil, just gas, or both.

Natural Gas

The primary ingredient of natural gas is methane, a gas created by the digestion process of microscopic organisms called archaea. These tiny organisms break down organic vegetable matter. They continue to provide this service in swamps, lake bottoms, termite stomachs, and the digestive systems of cows and other grazing animals. It was the gradual submersion and geological pressure on prehistoric swamps and vegetation that formed the deposits of coal, oil, and natural gas in the first place. The archaea produced the gas in these deposits while the organic solids became coal and oil.

try and now over half the nation's coal mining takes place west of the Mississippi.

The tapping of the low-sulfur coal deposits in the West has brought about a dramatic change in the coal industry. Wyoming currently supplies power plants in thirty-five states from the Powder River Basin. Eastern coal fields, located primarily in the densely forested hills of Appalachia and in Illinois, Indiana, and Ohio, were traditionally mined with underground techniques. The dramatic mountaintop removal style of coal mining now practiced in Appalachia is in direct response to the cheaper production methods used in Wyoming as the Eastern mining companies struggle to remain competitive. Only a third of U.S. coal reserves are located close enough to ground level to be surface mined. About two-thirds can only be accessed by underground mining. Growing demand for coal will mean finding new ways to access the deeper two-thirds of U.S. coal reserves, much of which is located in the more populous Eastern and Midwestern states.

Coal and the Railroad

Coal is usually transported by train, and the railroads have a long partnership with the coal industry. The two industries essentially grew up
together. Early trains in Britain ran on coal; indeed trains were initially
developed to haul coal from the mines in Wales. Because forests were so
dense and plentiful in America, wood was the first commonly used railroad fuel even for hauling carloads of coal into the cities. Coal was later
used to power locomotives, but that also changed when diesel trains
proved to be faster and easier to maintain.

Today electricity runs commuter trains but diesel still powers the
trains that haul the coal from Wyoming to electric generation plants in
the east. Because the railroad system has offered the most economical
means of transport for coal, both mining companies and power plants
that rely on railroad infrastructure are held hostage to the price of oil
when it rises. Coal may remain a plentiful domestic resource, but getting
it to the generation plants in populated areas of the country and in
the quantities they will continue to require already adds to the price of
"cheap" coal-fired electricity.

Using Coal

Historically speaking, coal has indeed gone through its own boom and
bust cycles. In about 1900, coal was providing most of the energy used in
the United States, but by 1950 it had been superseded by oil and natural
gas for many of its previous functions. This was the low point for coal
production in the United States, but its next boom was already on the
horizon. During World War II, the frenzy of manufacturing armaments
and military equipment to support the war effort drove the steady expansion of the electric utility industry. Added to that was the electrification of most of rural America. By the time the war was over and the
fifties era of prosperity began, the electric utilities industry was poised
for rapid and extensive growth. Even though nuclear power made its
debut about this time, coal became the primary fuel for generating electricity, which it remains to this day. About 92 percent of the nation's coal
production is currently used for that purpose. Just under 8 percent is
used in industry as coke for smelting iron ore and for manufacturing cement, and less than 1 percent is still used for heating.

Fig. 5. A Train hauling coal from the Black Thunder Mine in Wyoming. These trains are frequently as much as a mile long. Photo courtesy of The Center for Land Use Interpretation Photo Archive, Culver City, California

Coal Today—The Issues

The coal industry maintains that the main advantage of coal is that we have plenty of it right here in the United States, and this is true. The United States has the largest coal reserves in the world—13 percent of the nation's land has coal deposits under it. We even export about 2 percent of our coal production to countries that need it for their iron and steel industries. However, coal's environmental costs are high. Coal is the dirtiest of the fossil fuels to burn, producing more particulates, heavy metals, and other noxious substances than either oil or natural gas. It also produces the most carbon dioxide. Relying even more upon it in the future is difficult to justify as we face the grim uncertainties of climate change.

Carbon Dioxide

Carbon dioxide emitted from burning coal is its most serious atmospheric threat. Approximately 40 percent of America's CO_2 emissions are produced by the generation of electricity, primarily in coal-fired power plants. The second largest source is vehicle emissions at 33 percent. Essentially, the carbon dioxide stored underground in the ancient organic matter that became coal (or oil and natural gas) is released back into the atmosphere when it is burned. Carbon dioxide is necessary for plant life, but too much causes the greenhouse effect that can lead to global warming and climate change.

The U.S. Environmental Protection Agency (EPA) has not designated carbon dioxide and other greenhouse gasses to be pollutants, because of the federal government's long-standing claim that global warming has not been established to be a significant threat. By taking this position, the EPA has insisted it therefore does not have the authority to regulate it. That position was successfully challenged in April 2007 when the Supreme Court ruled in favor of the State of Massachusetts in its claim that CO_2 and three other nonregulated greenhouse gasses have caused actual and imminent harm, threatening its coastline from rising sea levels. The Supreme Court finding stated that "the harms associated with climate change are serious and well recognized" and that CO_2 fits the definition of an air pollutant as defined under the Clean Air Act. This court case, which focused on transportation emissions and involved twelve states and several local governments, has opened the door to expectations of serious federal regulation of greenhouse gas emissions from coal-fired power plants as well as vehicles in the not too distant future.

What the Government Does Regulate

However, the government has not ignored other atmospheric pollution, including substances emitted by the combustion of coal. In 1970 the EPA established acceptable air quality standards for what they call "criteria pollutants," namely ozone, carbon monoxide, lead, nitrogen dioxide (NO_X), sulfur dioxide (SO_2), and particle pollution (also called particulate matter), now categorized as coarse particulates or PM_{10}, which includes particles between 2.5 and 10 microns in diameter, and fine particulates or $PM_{2.5}$, which are 2.5 microns or less in diameter. NO_X, SO_2, and particulate matter are the three primary criteria pollutants produced at high levels when coal is burned.

Ozone

Ozone is produced primarily from motor vehicle exhaust, but it also comes from chemical solvents, industrial emissions, and coal-fired power plants. Ozone plays an interesting role in the earth's environment because it can be both a blessing and a curse. The ozone layer, located in the earth's upper atmosphere, protects us from the sun's ultraviolet rays. This is the good ozone, and we continue to work on reducing use of destructive chemicals in order to protect it. It is the ozone that forms at ground level that has been designated a criteria pollutant. Ground-level ozone results when a mixture of chemicals, including nitrogen oxide and a variety of volatile organic compounds, is exposed to sunlight. Ozone is the main component of smog and is known to aggravate asthma and to cause other respiratory problems. It also inhibits growth of vegetation by interfering with how plants and trees absorb carbon dioxide. The presence of high levels of ozone can reduce the effectiveness of planting trees and other vegetation to capture carbon dioxide.

Acid Rain

Coal-fired power plants produce 80 percent of the sulfur dioxide, and 30 percent of the nitrogen oxide that we emit to the atmosphere. These are the two primary ingredients of acid rain. Industrial processes produce much of the rest, and between 40 and 50 percent of NO_X comes from vehicle emissions. Acid rain happens when water droplets form around SO_2 and NO_X particles emitted from smokestacks and tailpipes and are carried long distances in the upper atmosphere. When the highly acidic droplets become heavy enough to fall as rain, they land in lakes and streams, raising the acidity of the water, which harms both surrounding vegetation and aquatic life. If soils are unable to neutralize the acid, it kills many plants and damages others. Damage in high altitude forests can also be caused directly by an acidic fog that lingers around trees and other vegetation, stripping nutrients from the leaves.

Emission clouds tend to move east with the general weather patterns causing damage sometimes hundreds of miles from the offending power plant. Eastern forests have been taking an acid bath in emissions from Midwestern power and industrial plants. The effort to combat acid rain was recognition at the federal level that pollution could and did cross state lines, making it a national problem requiring national solutions.

Acid rain also damages building materials and vehicle finishes, and erodes stone statues and monuments. Visibility is also affected by sulfates and nitrates. According to the EPA, sulfur dioxide causes 50 to 70 percent of the reduction in visibility resulting from pollution in the Northeast. The good news is that, since the implementation of the Clean Air Act Amendments of 1990, SO_2 emissions have been notably reduced, although NO_X emissions have remained about the same.

The Clean Air Act amendments of 1990 required utilities to meet strict emission levels of sulfur dioxide (SO_2) and nitrogen oxide (NO_X). Utilities could either install expensive equipment to trap these emissions or they could start burning coal with low sulfur content. Not surprisingly, switching to the relatively cleaner low-sulfur coal from Wyoming became the number one strategy among utilities all over the country for reducing their sulfur emissions.

Particulates Large and Small

Much particulate pollution is visible. A little digging into the history of coal use will show that people have objected to coal fumes and soot for hundreds of years. Queen Elizabeth I of England is said to have prohibited the burning of coal in London during sessions of Parliament to keep out-of-town members from becoming ill in the city air. London is famous for its fog, which was produced from the fumes of coal burned mostly to heat homes. But London is hardly alone. In the United States, the early centers of industry like Pittsburgh were cloaked in coal-fired industrial emissions, often creating darkness at noon, and in most cities, where coal was the standard residential heating fuel, both visibility and respiration were commonly affected. Now that we no longer use coal to heat our houses, and power plants capture the visible particulates from their stacks, we seldom think about that old-time soot and smoke.

But there is also particulate matter too small to see, which is formed when the emissions of burning fuels react with sunlight and water vapor. All particulates under 10 microns (PM_{10}) can be inhaled into the lungs, causing a variety of respiratory conditions. However, recent research is finding that the fine particulates ($PM_{2.5}$) are the most dangerous because they are small enough to become much more deeply embedded in the lungs. One person in three is at risk from complications caused by $PM_{2.5}$, particularly those with heart or lung conditions, children, the elderly, or anyone who is active outdoors. In 2006 the federal government revised the 1997 standards for $PM_{2.5}$, but designation of areas out of compliance with these new levels will not be established until 2010, and states and local governments will have three years to comply with the regulations. Even with regulation in place it is likely to be a considerable technical challenge to capture fine particulates in sufficient amounts to substantially lower the risk, and it is also likely to be expensive.

EPA Attainment Areas

EPA-designated geographic areas are either "in attainment" or "non-attainment" depending on the levels of the criteria pollutants in the air. The regulations are designed so that levels can be adjusted as we learn more about the effects of these pollutants. In 1997 the EPA proposed new standards for particulate matter and ground level ozone (primarily from automobile exhaust), and new concerns about the health effects of fine particulates may tighten these levels further. While the criteria pollutants are certainly not the only toxic components of fossil fuel combustion, their status as primary health and environmental hazards is well founded in scientific evidence. The regulatory process for pollutants is highly controversial because of the economic implications for those who must install the necessary and frequently expensive equipment to capture them. Therefore, bringing a pollutant under tighter regulation can be a long and arduous process. However, federal regulation has been successful in reducing some criteria pollutants, particularly sulfur dioxide.

And Then There Is Mercury

The Clean Air Act also regulates 188 hazardous air pollutants, of which mercury is one. Coal-fired power generation is the greatest source of mercury from human activity. Mercury is an elusive and complex element because many of the problems it causes are the result of other chemicals and processes interacting with it. While mercury can take a variety of forms, the most worrying is methyl mercury, which is classified as a neurotoxin, or a poison that attacks the nervous system. Exposure to methyl mercury in the United States is primarily attributed to coal-fired power plant emissions that leave the smokestack as airborne particles containing mercury and fall on the landscape. Some mercury falls directly into lakes, streams, and rivers, and some works its way into water bodies through the watershed. Certain species of bacteria in lakebed sediment ingest the mercury particles, transforming it to the toxic form, methyl mercury, and then themselves become the food source for larger creatures and fish. All the methyl mercury accumulates through the food chain as progressively larger creatures are eaten, until the fish that people catch to eat—such as bass, pike, walleye, or brown trout in fresh waters, or ocean fish species including tuna, swordfish, and halibut—end up containing dangerous levels of methyl mercury. Almost all states now have fish advisories that suggest safe levels of fish consumption, particularly for young women and children. Most states consider methyl mercury contamination of lakes to be a serious public health issue, and many also regard it as dangerous to the well-being of wildlife and natural systems.

Mercury is a controversial pollutant not because its health effects are debated but because so much of it occurs naturally and it is impossible to tell whether the mercury in any given lake came from the local soil, from a power plant, or from some other source. Also, emissions from power plants and other industries can travel great distances, so sources of man-made mercury pollution are hard to identify precisely. Because of this complex behavior, mercury is politically challenging to regulate. In 2005, the EPA issued the Clean Air Mercury Rule, which was the first federal attempt at mercury regulation. This rule would essentially reduce mercury emissions 69 percent by 2018. However, over twenty states have already adopted stricter standards, with some reduction goals as high as 80 or 90 percent, and they require compliance sooner than the federal regulations. The federal government has chosen to agree with the position put forth by the utility industry that effective controls are not yet economical for power stations to install, while the emerging state regulations reflect greater confidence that the presence of stricter laws will encourage development of the technology required.

Environmental Impacts of the Coal Industry

The mining of coal contributes to environmental degradation in other ways, particularly destruction of habitat and the poisoning of ground water. Mining operations produce considerable toxic waste material from both the extraction and processing of coal, and these toxins can leach into the ground water. Mountaintop removal mining in Appalachia disrupts watersheds when topsoil and debris are deposited in valleys where streams and rivers run, destroy vegetation that anchors the hillsides, and cause landslides and flooding. While reclamation practices can be efficient and economical on the flat and unpopulated prairies of Wyoming, they are entirely inadequate for returning the forests and communities of Appalachia to their former functionality.

Coal Industry—Social Issues Continue

The coal industry has been both politically and socially controversial as an employer. Typical mining operations and "company towns" in the United States, Britain, and elsewhere have traditionally been isolated communities, with the owner or manager having considerable power over the workers' lives both on and off the job. Because of this isolation and the potential for the abuse of managerial power, it is not surprising that the working and living conditions in mining towns sparked powerful union movements in the early twentieth century. Miners were primarily concerned with safety issues and working conditions. Coal mining is dangerous work in the best of circumstances, and the industry has a history of cutting corners regardless of known risks to its employees. In addition to the ever-present perils of cave-ins and explosions, many miners fell victim to the slow death of black lung and other occupational diseases. Even though working conditions have improved over the years, it is still a tough way to make a living. Today, employment in coal mining is in decline. The high-sulfur bituminous mines in the East, primarily the labor-intensive underground operations, have lost out to the more economical Western surface-mine operations that require fewer workers.

Considering the pay scale and the dangers, coal mining is not regarded as a promising career opportunity by young workers in the United States. This is evidenced by the fact that the average age of coal miners still working is well past forty. The next employment trend in U.S. coal mines might be the importing of experienced workers from other countries willing to accept the comparatively high pay and safer conditions of U.S. mining operations, particularly if the industry can recrown itself king of U.S. energy, a position it is actively seeking.

Coal—What's Next?

The coal industry remains optimistic about increasing its market share because coal is abundant within our borders and therefore quite attractive politically. The industry is promoting two technologies in particular: converting coal to liquid fuels, and clean coal technologies for electric generation. With coal, they argue, we can solve our transportation problems with liquid fuel from coal and our rising demand for electricity through "clean coal" generation technologies, thereby reducing our dependence on foreign energy. However, both these ambitious goals depend on our developing the ability to "sequester" the carbon that would be released.

Carbon Sequestration

To sequester carbon means to capture and store carbon dioxide, thereby removing it from the atmosphere where it would exacerbate global warming. The CO_2 we are now releasing by burning coal, oil, and natural gas was originally sequestered by the ancient swamp plants and marine organisms that decomposed under pressure and has actually been stored for millions of years underground. Much of this extra CO_2 is already soaked up by the oceans and the vegetation in the world's rain forests. The oceans are able to sequester about a third of human-generated CO_2 production. But this natural carbon storage can't keep up with the amount of CO_2 we are now producing through fossil fuel combustion, which is why we need to reestablish the balance either by burning considerably less of these fuels or finding ways to artificially sequester CO_2.

We could plant more trees and other vegetation to increase natural sequestration. For years, efforts have been made to preserve and extend forest lands to protect habitat and wildness. These efforts have now taken on a new urgency, particularly in rain forest areas around the world. Both forestry and cropland management are recognized as important for control of greenhouse gasses, not only for CO_2 but also for nitrogen oxide and methane. The EPA has identified several ways that forestry and agriculture can assist in this process.[1] Besides simply planting more trees, they suggest preserving forest lands and enhancing the growth of existing trees. They also recommend agricultural practices such as conservation tillage, which reduces disturbance of the soil, and rotating the location of grazing livestock to improve forage, thereby reducing production of methane and nitrogen oxide. Agriculture also has the potential to produce new biofuels that are considered to be net "carbon neutral" because, when burned, they emit the CO_2 they pulled from the atmosphere, which is absorbed in turn by the next crop in the fields.

Sequestration Technologies

Besides natural sequestration techniques, there are technological approaches, such as collecting the CO_2 at some point in the combustion process, liquefying it, and sending it through pipelines into storage. Carbon is most easily captured at sources of high CO_2 production such as power plants, natural gas refineries, and plants where hydrogen is made from natural gas. Collecting it at a precombustion stage by gasifying the coal is far more economical than attempting to capture it from the smokestack. It is very difficult to collect atmospheric CO_2 emitted from vehicles and diverse small sources.

The strategy of capturing and storing CO_2 in great quantities as a future solution to global warming is a popular mantra for politicians, but the hoopla may be a bit before its time. First, artificial sequestration processes are energy-intensive themselves, which will automatically increase the cost of the electricity as well as reduce the overall efficiency of the energy use. Second, once captured the CO_2 must be stored securely. Injecting CO_2 into gas and oil fields, marginal coal seams, and underground saline aquifers or brine formations is being considered. It is also possible to store it in caverns under the ocean. The first undersea industrial-scale CO_2 storage is located one kilometer below the seabed off the coast of Norway.

Some methods, like enhanced oil recovery, actually use the CO_2 injections to force more oil to the surface, creating an additional product from the process, but location would be important in making this economical. Obviously, it is necessary to find safe and permanent storage locations, but little is known yet about the environmental implications of any of these strategies. These techniques are all being considered in view of the massive quantities of CO_2 that could be captured in the future. Location of potential storage sites, pipelines, and other infrastructure, along with a system of regulatory oversight, will be required before such quantities could be handled. It could easily be another twenty years before we see such sequestration infrastructure evolve.

Commercial Carbon Dioxide

A significant volume of pure CO_2 is already produced in the United States for industrial purposes. CO_2 is used to manufacture a variety of products, including furniture polish, hair spray, automotive chemicals, and deodorant, and is the chief ingredient in a number of pesticides used primarily to control insects and mites. In its solid form called dry ice, it transforms directly from a solid to a gas. In this form it is used as pellets for blast cleaning industrial equipment, for carbonating beverages, and for transporting cargo that must remain cold. When placed in water its transformation to gas is accelerated and it produces a dense and dramatic fog, often used to good effect on the theatrical stage. CO_2 does not maintain a liquid state at normal atmospheric pressure, which means that sequestering and storing it as a liquid requires pressurization.

Commercial CO_2 is produced in a variety of ways, usually as the byproduct of another fossil fuel refinement process. However, the commercial supply is not yet coming from sources where carbon sequestration is the primary objective. The challenge for clean coal electricity generation will be removing the CO_2 economically under high volume circumstances, focusing on quantity rather than quality. Commercial uses of this CO_2 might include enhanced oil recovery, or potentially recovery of methane from coal seams that are not accessible for mining, although the latter has not yet become economical.

Clean Coal Technologies

Clean coal technology is not a new idea, nor does it address all the pollution issues associated with burning coal. The U.S. Department of Energy coined the term when it established the Clean Coal Technology Program in 1986 to reduce emissions of NO_x and SO_2, the components of acid rain, and also to improve the overall efficiency of coal combustion technologies. The program partnered U.S. DOE with state governments and the coal industry, which provided 66 percent of the $5.3 billion for 38 demonstration projects in 18 states. These projects were chosen to demonstrate technologies that could become commercially viable, and the partnership was a happy one. The $1.8 billion in federal funding was the first large monetary investment the government made in reducing the environmental impacts of coal. The participating states were interested in hosting demonstration projects that would eventually boost the productivity of their local coal economy. The coal industry itself, aware of impending regulations, was pleased to get some federal funding to help them comply.

The resulting technologies addressed pollutant capture at three points in the combustion process—before the coal is burned, during the combustion process, and after combustion when the emissions are leaving the stack. Much of this technology continues to focus on removing sulfur and lowering NO_X emissions. Physically cleaning the coal before combustion can be done simply by washing it. This can remove up to 90 percent of the sulfur and reduce SO_2 and ash. There is also a chemical process called molten caustic leaching that removes sulfur and other minerals. New methods under development use sulfur-eating bacteria to munch away the sulfur in the coal.

Low NO_X burners use a technique called fluidized bed combustion, which suspends a mixture of pulverized coal and limestone on jets of air, moving the mixture like a fluid and allowing the limestone to soak up the sulfur in the coal. Not only is sulfur reduced, this process needs less heat than the traditional pulverized feed method, so less NO_X is produced. This process also lowers NO_X emissions because it requires lower temperatures to operate, meaning less NO_X is formed. Also, fluidized-bed combustion has an additional feature that can increase combustion efficiency. The fluidized bed of burning coal heats water to make steam for a steam turbine electricity generator. Then, the resulting flue gasses, which have been substantially cleaned by the limestone in the process, can be burned to power a secondary gas turbine generator to make additional electricity.

Flue gas desulfurization or "scrubbers" had been used on power plant stacks since 1967. Scrubbers mix the flue gasses with a slurry of limestone to soak up the sulfur. A byproduct of this process is gypsum, which is recycled and used to make sheetrock. Another technology in common use is the electrostatic precipitator, which uses static electricity to remove fly ash and large particulates from the flue gas. Post-combustion scrubbers and precipitators remove mercury emissions to some extent as well. Mercury is more difficult to control because there are so many variables in the characteristics of coal and the design of power plants. The most promising technology for mercury removal is a process called activated carbon injection (ACI), which is already in use for commercial waste incinerators. In this process, powdered, activated carbon is injected into the flue gasses at a point just before the particulate filter. The carbon particles bind with the mercury and are then caught in the filter. It will be necessary to adapt this technology for power plants because they differ from incinerators both in scale and in the chemical content of their flue gasses.

All of these methods had contributed something to improving the cleanliness of coal combustion, but the efficiency of coal plants had not been significantly improved over the 30 to 35 percent efficiency achieved in the 1950s and 1960s. In contrast, coal plant efficiency in the 1900s was only 5 percent. The Clean Coal Technology Program claimed success in 22 of its 38 demonstration projects, even though a major leap in efficiency has still not occurred. In 2001, George Bush announced $2 billion in funding for the Clean Coal Power Initiative, a second round of ambitious demonstration projects. This time the list of objectives once again included reducing SO_2 and NO_X, but it added the capture of mercury and carbon dioxide emissions, as well as the improvement of combustion efficiency in order to use less coal.

None of the technologies mentioned so far are designed to capture CO_2. The clean coal innovations that have captured recent media attention address the big carbon question about how to separate the CO_2 in such a way that it can be easily diverted to a pipeline for storage or use.

FutureGen

The U.S. DOE list of projects in the Clean Coal Power Initiative includes FutureGen, a $1.77 billion generation plant initially undertaken as a public-private partnership. FutureGen is George Bush's presidential initiative to build the first zero emissions coal plant in the world. Its design intends to integrate hydrogen production for power generation with sequestration of the CO_2 from the coal, to generate about 275 megawatts of power. The net efficiency is estimated to be 40 percent, not a particularly notable improvement on the average coal plant efficiency of 35 percent. FutureGen's purpose is to demonstrate that fossil fuels can be used in a carbon constrained world, and it clearly underscores the federal government's commitment to continue supporting fossil fuel technologies as the primary energy path for the future, regardless of cost.

FutureGen's price tag of $1.77 billion seems alarmingly high even for a demonstration project. Assuming that actual construction does not exceed the estimated budget, the cost to build it will be a hefty $6.49 million per megawatt of power. A typical new coal plant using the standard pulverized coal technology with emission controls that meet current U.S. EPA standards can be built for less than $2 million per megawatt of power. However, such a plant, typically built to produce 600 megawatts of power, will pump four to five million tons of CO_2 into the atmosphere every year. Assuming that its carbon emissions can be sequestered along with the other pollutants, the "zero" in "zero emissions" from the FutureGen plant will cost close to $4.5 million per megawatt of power. And that price only includes removing the CO_2 from the combustion process. It does not include the cost of safely storing it underground or elsewhere. Even if that "zero emissions" premium could be cut in half when the technology is commercialized, it still illustrates how expensive clean coal technology may turn out to be. In early 2008, the FutureGen project suffered a serious hitch when the U.S. DOE withdrew its funding commitment, claiming concern with rising costs. The FutureGen Alliance, the private industry partner, will attempt to move forward, but the project is seriously delayed.

For comparison we can look at the cost of wind energy, another domestic energy resource, which is frequently criticized for its up-front costs. Depending on the site, wind farm installation currently costs about $1.5 million to $2 million per megawatt of power, which is competitive with a traditional coal-fired plant at $2 millon per megawatt. Wind energy costs are projected to decrease as the U.S. turbine manufacturing industry grows and equipment prices come down. Even though the wind does not blow everywhere all the time, the United States has significant wind power potential that could be economically tapped. Wind turbines produce no greenhouse gases and their maintenance cost is low.

Wind power, solar energy, and other renewable resources are usually dismissed as being too expensive to contribute to reducing greenhouse gas emissions any time soon. It's true they cost more per megawatt than standard coal-fired plants at the present time, but the advanced clean coal technologies currently in development can hardly claim to be more cost-effective. Widespread adoption of clean coal technologies will change the economics of coal-fired power dramatically. Cleaning up coal requires highly complex technical solutions that will mean new coal plants will be expensive to build and maintain, and we will still be paying for the extraction impacts of mining. Whatever we do to provide ourselves with electricity in the future, there is no question costs will increase.

Cleaning Coal Through Gasification

Gasifying coal breaks up its chemical structure so that its components can be selectively used or removed. In a coal gasification process, coal is exposed under pressure to heat and steam, along with a controlled measure of oxygen. This breaks the coal down into its component elements and compounds, primarily hydrogen, carbon monoxide, and carbon dioxide. These three components comprise "syngas" (or synthesis gas), which can then be burned. Syngas can be refined further to remove or separate other chemical impurities such as sulfur and mercury. Gasification can remove over 90 percent of criteria pollutants and mercury before combustion. The purified syngas can then be burned with few emissions. The impurities removed in the gasification process can sometimes be sold for other uses like fertilizer or cement through existing markets. While CO_2 can be efficiently captured in the gasification process, storing it requires a safe and secure location where it can truly be "sequestered."

We are not without precedent for demonstrations of sequestering the carbon extracted from gasified coal. One gasification plant, something of an anomaly, was the first in North America to capture its CO_2 and sell it. The Great Plains Synfuels Plant in North Dakota, owned by Basin Electric Power Cooperative, sequesters CO_2 from a coal gasification process that produces synthetic natural gas from lignite coal. The plant was built in the 1970s in response to fears that the oil crisis would reduce the supply of natural gas as well. Up until the 1990s, the plant did not sequester its CO_2, but simply released it into the atmosphere. When Basin Electric found they could sell their CO_2 to a petroleum producer 200 miles north in Canada to use for enhanced oil recovery, they refitted the plant to add carbon capture to the process and built a pipeline to transport it. Capturing CO_2 as part of a coal gasifying process is relatively economical when compared to capture during combustion or post-combustion.

The gasification process at Great Plains distills some other profitable products from the gas stream as well, including anhydrous ammonia and ammonium sulfate, which can be sold as agricultural fertilizers.

Integrated Gasification Combined Cycle or IGCC

Gasification is truly an advanced technology because the coal itself is not burned, but rather it is transformed into a purified gas that is then burned. Not only can pollutants be isolated before combustion, the potential efficiency increases by combining the syngas with a combined cycle, or two-turbine system for generating electricity. The technology currently considered most likely to succeed is called Integrated Gasification Combined Cycle or IGCC. In this system, syngas consisting of hydrogen and carbon is produced and burned to run a gas turbine to generate electricity. The heat from this combustion process is then captured to make steam that powers a second, steam turbine to generate additional electricity. Efficiency for IGCC systems is generally assumed to be about 40 percent, similar to the FutureGen project.

There are two IGCC plants up and running in the United States. One is the 260-megawatt Wabash River Station near Terra Haute, Indiana, which began operating in 1995. Wabash River Station was one of the projects supported by the original Clean Coal Technology Program to demonstrate the repowering of an existing coal-fired plant with IGCC technology. Duke Energy is planning a second IGCC plant in Edwardsport, Indiana, which will replace two aging coal-fired plants. Neither of these IGCC plants will be capturing CO_2. The second IGCC plant already in operation is Tampa Electric's 260-megawatt Polk Power Plant located near Tampa, Florida. While this system removes most of the criteria pollutants and mercury, and recycles its cooling water, it isn't capturing CO_2 either.

The technology sounds truly miraculous, so why are there not more IGCC plants being built? The main reason is that IGCC is still experimental and very expensive, and therefore too risky for private investors, even those with deep pockets. Furthermore, utilities have been able to meet regulatory requirements in other ways. Many utilities put their new infrastructure money into natural gas turbines during the 1990s when natural gas was cheap. Also, a majority of U.S. power plants now burn lower sulfur coal from Western mines to meet the regulatory limits. And then there are the old, coal-fired plants, grandfathered in without emissions controls, which have been kept on line past their retirement age. Significant upgrades to these obsolete plants would mean they would have to meet current regulatory emissions standards, which would require a lot of expensive new equipment and lose them their grandfather status.

Utilities are staying just ahead of the game with the currently regulated pollutants, but when the federal government begins regulating CO_2 and other greenhouse gasses, many dramatic changes are in store for both the coal and the utility industries. There are already signs that investors are realizing the rapidly diminishing returns on an investment in an expensive coal plant that might not meet future regulations. Sequestering carbon as part of coal-powered electricity generation will require a whole new energy infrastructure.

Coal to Liquid Transportation Fuels

Coal to liquid, or CTL fuel for transportation has been getting a lot of attention politically. The pressure is tremendous to find something other than petroleum to put in our gas tanks. Ethanol and other biofuels also have their proponents in what has become a desperate competition to reduce our dependence on foreign oil. But the development of biofuels at the scale we will require is just beginning, and the industry is experiencing the difficulties of infancy like any young venture as it searches for sources and technologies. CTL proponents claim the coal industry already knows where its resource is coming from and has the infrastructure in place to extract it. Also, the technology to produce liquid fuel from coal has been around for a long time, having been developed by the Germans during World War II. Politicians from both sides of the aisle, particularly those from coal-rich states, consider CTL fuel to be a promising idea.

However, in our rush to use coal for reclaiming domestic production of transportation fuel, we risk increasing the current environmental decline caused by using foreign petroleum. Unfortunately, CTL fuel ultimately produces about twice the carbon dioxide as gasoline when taking into account both its manufacture and use. And even though a number of CTL plants are already in the planning stage, development of the technologies promised for sequestering the carbon produced from the fuel's manufacture has not yet happened. Furthermore, because the

U.S. EPA does not yet regulate CO_2 emissions, CTL projects are being proposed without budgeting for the costs of carbon sequestration, which will surely increase the price tag once regulations are in place. Still, according to the National Energy Technology Laboratory, plants in Illinois, Pennsylvania, and Wyoming will be up and running by 2009, without sequestration technologies in place. The environmental impacts of using coal for transportation fuel could be enormous, and the costs of mitigating those impacts could make both taxpayers and investors leery.

Oil: King of the Road

Known technically as petroleum (Latin for "rock oil"), oil has taken the lead in U.S. fossil fuel consumption, primarily because we use it for transportation. This includes not only Priuses, Cadillacs, Volkswagens, minivans, Hummers, and pickup trucks, but also the tractor-trailers that carry food to the supermarkets, toys to Walmart, and machine parts to factories. Petroleum runs the trains that haul coal to power stations, the boats and ships at sea, and aircraft, from single-engine Cessnas to 747s and F-16s to the space shuttle. Petroleum keeps both us and our economy moving. Petroleum is also an essential raw material for the production of an astounding variety of products, many of which are made from the byproducts of refining petroleum to produce gasoline, propane, and kerosene. Plastics, synthetic fabrics, cosmetics, household cleaners, and medicines are a few.

Oil's potential was initially misunderstood. Oil seeps would occasionally surface but there was little use for this black liquid, although Native Americans used it to treat frostbite, and it enjoyed some popularity among European Americans in the early nineteenth century as a patent medicine called Seneca Oil or Rock Oil. At about this time, water drillers would occasionally bring up oil with their efforts, but they considered it a nuisance because it contaminated the water. However, its value suddenly increased during the decade between 1850 and 1860. First, in 1849, Canadian Geologist Abraham Gesner developed a process for distilling kerosene from petroleum. In 1854, the Pennsylvania Rock Oil Company was founded as the first U.S. oil company. Then, a commercially viable kerosene lamp was invented in 1857, and in 1859 the first commercial oil well was drilled in the United States at Titusville, Pennsylvania. Kerosene lamps soon replaced whale oil lamps, and the oil industry was born. Kerosene was the primary product of the oil industry until the 1920s when gasoline for automobiles took the lead.

Saving the Whales and the Horses

As an ironic side note, considering the poor reputation oil presently has for its environmental impacts, the adoption of kerosene as a substitute for whale oil is credited by many for saving several species of whales, including sperm whales and right whales from extinction. Unlike candles made from tallow or lamp oils made from animal fats, whale oil burned with little odor or smoke and was therefore preferred by those who could afford it. A huge whaling industry had developed by the 1850s producing between four and five million gallons of oil annually. If kerosene had not been developed and marketed for lamp fuel there is little doubt we would have lost several species of whales altogether.

Oil was later partially responsible for relieving another environmental dilemma as well when the automobile and the electric streetcar replaced the horse for urban transportation. As the 1900s approached, large cities like New York and Chicago faced a significant pollution problem from horse manure and carcasses. Horses used for both private and public transport produced great quantities of urine and manure in their daily routes. It made for unpleasant walking, and the flies created a huge health hazard. Horses, often mistreated and overworked, would frequently die in the streets. It was the hard life of the urban horse that inspired Henry Bergh to found the American Society for the Prevention of Cruelty to Animals in 1866. City governments were charged with removing dead horses, which were dumped into landfills or taken to highly odorous rendering plants that no one wanted in their backyard. In 1880, 15,000 dead horses were removed from the streets of New York City. An additional complaint from urbanites about horses was the noise they made with their steel shoes and the sound of wagon and carriage wheels on the cobblestones. By 1888, most major cities had electrified their streetcars, and other vehicles like taxicabs and busses became motorized. By 1912, there were more motor vehicles than horse-drawn conveyances in New York, London and Paris.[2] Automobiles, which had previously been banned in many cities because it was feared they would cause traffic jams, became the solution to equine pollution.

Oil: Today's Issues

Automobiles and other motor traffic did go on to cause traffic jams, but the rapid growth in gasoline consumption during the twentieth century has created its own detrimental environmental impacts as well. Petroleum transportation fuels, primarily gasoline, diesel, and aviation fuels, contribute half the nation's NO_X emissions and 38 percent of the CO_2. Even though oil produces half the CO_2 emissions that coal does, we use about twice as much oil as coal. Coal combustion from power plants produces about 40 percent of our CO_2 emissions. Transportation fuels also emit a toxic mixture of other pollutants as well, like carbon monoxide, ozone, methane, particulates, and SO_2. The resulting cocktail of emissions produces acid rain and smog, and heightens global warming.

The fuels themselves are also pollutants, causing considerable damage during shipping and refining. According to the Union of Concerned Scientists, the oil shipping industry floats 6,600 oil tankers that move 524 billion gallons per year; 230 million gallons of oil were spilled in U.S. waters between 1973 and 1993, which works out to an average of 31,000 gallons of oil every day for twenty years.[3] Tanker spills are not even the greatest source of petroleum contamination to reach water and soil. Leaks from pipes and tanks at refineries, gas stations, and other fuel storage facilities, dumped waste oil, dripping crankcases, and pavement runoff all add up to over 85 percent of the pollution from oil.

Controlling carbon emissions in the transportation sector requires a solution very different from a technology like coal gasification where the CO_2 is removed before the fuel is burned. Controlling carbon emissions from millions of tailpipes would be impossible, so carbon reduction strategies for vehicles tend toward superefficiency or alternate fuels, including biofuels like ethanol and biodiesel, and hydrogen fuel cells. In other words, reducing carbon emissions from vehicles will mean actually reducing the use of petroleum, rather than trying to clean it up like coal for electricity generation.

Dwindling Supplies, or "Peak Oil"

While the environmental impacts of our petroleum use urgently require our attention as part of the whole global warming issue, more urgent still is dealing with the fact that we are already using it faster than we can produce it domestically, and that will apply internationally as well, some say within the next decade. The point at which this reversal occurs is called the "peak" of the oil supply.

Because of "peak oil," Many experts anticipate sharp and permanent increases in oil prices that will likely bring a severe chill to the U.S. economy. It's not simply a question of supporting the personal transportation patterns built into our suburban culture, but the whole structure of commerce is imperiled as well. Everything from supermarkets and big box stores to the construction industry and FedEx rely on transportation that runs on oil. Prices for food, hard goods, housing, and all kinds of services would rise, and demand would fall as people tighten their belts. Even though environmental concerns are a primary theme, the campaigns promoting support for local farmers, manufacturers, and sources of building materials and other products are also directly related to confronting the challenge of reducing the need for transportation fuel.

As world demand for oil rises with the rapid development of economies like China and India, competition will put pressure on everything from international political and economic stability to preserving what's left of the natural environment, particularly where dwindling supplies of oil can be found. U.S. competition with developing nations for world supplies of oil and natural gas is made more complicated because much of remaining reserves of both are owned by nations not necessarily friendly to the United States or other Western nations.

Between the optimists and the pessimists, there is a wide range of opinion on exactly when peak oil will take place, but most experts recognize that it is close enough to begin having definite impacts on the world's economy. In the past there has been a comfortable cushion of reserves internationally to even out the bumps and blips in the day-to-day supply. Saudi Arabia in particular has provided a backup for the world's supply during supply emergencies like the first Gulf War when oil supplies from both Kuwait and Iraq were cut off. Even though Saudi Arabia reveals little about its reserves, there are industry observers who believe it no longer has the capacity to act as the world's oil bank, protecting the world's consumers from the severe weather events or political upheaval in key production locations that can disrupt the world supply, causing prices to fluctuate wildly. The United States is particularly vulnerable to such fluctuations because of our economy's tremendous dependence on petroleum. Obviously, anything we can do to use less petroleum would be to our benefit, and it will likely require a mixture of strategies including bio-fuels, ultra-high-efficiency and electric vehicles, increased public trans-portation, and higher density urban design, and using business electron-ics more efficiently to reduce commuting and other professional travel.

Finding the Oil That's Still There

The peak oil phenonemon doesn't mean that the world's oil is close to being used up. There is still plenty of oil underground, but much of what's left is not easily accessible, or accessible without potentially sub-stantial environmental impact. We continue to search for new oil fields, and prospecting technology has improved. Oil engineers use computer-ized seismic analysis by setting off surface explosions and tracking the patterns of the sound waves as they travel through the ground. What they're finding, however, are smaller and more scattered conventional oil deposits than they used to find. Extracting this oil will require more oil crews spread over greater areas to produce the same or less oil. Un-conventional oil deposits, like tar sands or oil shale will require us to use even more laborious and energy-intensive production methods. Reach-ing the "peak" really means that we have reached the point where we will have to work much harder and longer to produce the same amount of oil. It is a problem of production capacity more than one of overall sup-ply. A brief look at conventional and unconventional recovery methods illustrates this point.

Oil does not occur in liquid pools or pockets that can be tapped with something resembling a soda straw as is commonly believed. Oil is lighter than water, so it tends to migrate upward from its sedimentary begin-nings and becomes trapped in porous rock formations on the way to the surface. Its progress can be blocked by layers of nonporous stone, and over the years pressure builds. With conventional oil recovery, prospec-tors look for dome-shaped formations of nonporous stone, as these can produce the famous "gushers" where oil is pushed out under its own pressure. As an oil well ages, the pressure within it diminishes and this is when "enhanced oil recovery" measures are used, primarily injecting water, natural gas, or potentially liquid CO_2 into the well to increase the underground pressure once again to keep pushing up the remaining oil.

Conventional recovery can only reach a small portion of the world's reserves because most oil resides in tar sands. Tar sand oil is referred to as "unconventional oil" because it requires different, more expensive recovery methods than those currently used. The largest deposits of tar sands are in Alberta, Canada, and Venezuela, although there are some small deposits in Utah. Canada is commercially extracting more than a million barrels of oil a day from its Alberta tar sand deposits in large, open pit mining operations. Canadian tar sands oil extraction constitutes 40 percent of their oil production. The oil in tar sands is in the form of bitumen, a very heavy oil, which occurs in mixture with sand, clay, and water. After the sands are scraped from large, open pit mines, the bitumen is separated with a hot water process and then refined into oil. It takes two tons of tar sands and several barrels of water to produce one 42 gallon barrel of oil. The rising price of oil may soon make tar sands recovery economical in Utah, but the fields there are on public land and there are significant environmental impacts to consider. For local residents, the amount of water required for the process will be the greatest concern.

Another source of unconventional oil is oil shale. The United States does have some of the world's largest deposits of oil shale, mainly located near the Green River in Colorado, Utah, and Wyoming. The Green River Formation, most of which is located on public land, is very tempting to the oil industry because it is estimated to contain at least three times the known oil resources of Saudi Arabia. Oil shale extraction technology has yet to be used at commercial scale because costs are significantly higher than conventional recovery of oil. However, as prices rise and unconventional sources become more economical, the oil industry will find it worthwhile to develop more efficient recovery technologies for unconventional oil.

Oil shale is a rock that contains quantities of kerogen, the substance that, given time, heat, and pressure would turn into oil. Removing the kerogen from the shale requires a high temperature process called retorting, where the shale is heated in a vessel called a retort. Oil shale can be extracted from either underground or surface mines, or there is a process under development that heats the shale where it lies underground over a period of several years, where it eventually releases the oil from the rock, and it can then be conventionally recovered.

For the United States to satisfy its 20.7 million barrel daily demand for oil from tar sands or oil shale, it would have to be producing over twenty times what Canada presently does. Neither tar sands nor oil shale are

What's in a Barrel of Oil?

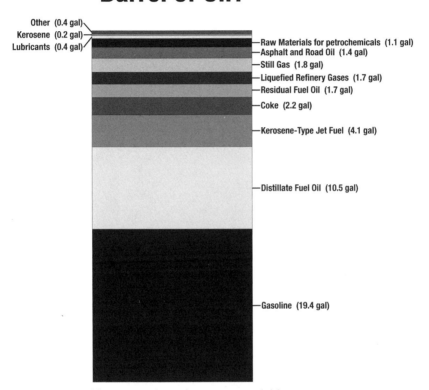

Other (0.4 gal)
Kerosene (0.2 gal)
Lubricants (0.4 gal)
Raw Materials for petrochemicals (1.1 gal)
Asphalt and Road Oil (1.4 gal)
Still Gas (1.8 gal)
Liquefied Refinery Gases (1.7 gal)
Residual Fuel Oil (1.7 gal)
Coke (2.2 gal)
Kerosene-Type Jet Fuel (4.1 gal)
Distillate Fuel Oil (10.5 gal)
Gasoline (19.4 gal)

Figures are based on average yield for U.S refineries in 2005. One barrel contains 42 gallons of crude oil. The total volume of products made is 2.7 gallons greater than the original volume. This represents "processing gain."

Fig. 6. What's in a Barrel of Oil? Source of data: American Petroleum Institute

poised to provide notable quantities of petroleum any time soon. With conventional oil harder to come by, and these unconventional sources not yet ready for prime time, we are rapidly finding ourselves in the place where supply meets demand, and there is no wiggle room. A number of other factors are also emerging, like rapidly rising demand from the new, high-volume petroleum customers India and China, and with escalating environmental concern over greenhouse gas emissions. In the United States, the peak oil and environmental issues are beginning to overlap and conflict with one another. Results from a survey by the Pew Research Center released in September 2005 strongly suggest that as the price of oil rises, so does American willingness to drill for oil in the Arctic National Wildlife Refuge.[4]

Our Petroleum Byproduct Culture

America loves to drive, and we also love our computers, jet-skis, cosmetics, Game Boys, and bottled water. Many people don't realize that our need for petroleum goes well beyond transportation. We don't import crude oil from Saudi Arabia to make lipstick or bowling balls. We get these things as a result of needing to fill our gas tanks. Because we use so much transportation fuel, it has become extremely economical to rely on petroleum as the raw material for a vast number of other things. Most of the oil in a barrel goes for gasoline, fuel oil, jet fuel, and other fuels and lubricants, but about 3 percent becomes petrochemical feedstocks.

The chemistry is complex but the idea is simple—make use of the byproducts of fossil fuel refining by breaking them down into the elemental building blocks to make a wide variety of synthetic products, producing them more cheaply than the naturally based products they are replacing. Polyester and nylon are cheaper and easier to produce than cotton and wool. Plastic automobile components are cheaper than metal ones and cost less to transport. If we didn't have plastic packaging and synthetic preservatives, would we be able to have the inventory control and shelf life necessary for operating supermarkets and big box stores economically?

Because coal, oil, and natural gas are chemically similar, it is possible to make many of the same things from any of them by breaking them down into their component elements and compounds. The concept of finding commercial uses for fossil fuel byproducts began in the early

Some Components of Petroleum	
Benzene • Synthetic rubber • Dyes and inks • Plastics	**Xylene** • Industrial solvent • Paint thinner • Perfumes and cosmetics
Toluene • Paint thinner • Solvents • Adhesives • Fingernail polish	**Phenols** • Resins • Nylon and other synthetic fibers • Disinfectants and germicides

nineteenth century with coal. The production of high-carbon coke for smelting iron ore involves baking coal in high temperatures to vaporize impurities. One valuable byproduct was coal gas or manufactured gas, which later developed into a whole lighting fuel industry of its own.

Another byproduct was coal tar, a thick and viscous substance that was not as obviously useful. Coal tar was well on the way to becoming a major pollution problem until the 1850s when chemists began finding ways to separate and use its organic chemical components. Among them are benzene, toluene, naphthalene, xylene, and phenols, which are now essential for making everything from drugs and plastics to explosives, paints, and perfumes. Among the first products developed from coal tar were aspirin and aniline dye. The first plastic from coal tar, called Bakelite, was developed in 1910. The separation of chemical raw materials from coal tar marked the beginning of all the synthetic products we use today. These chemicals can also be distilled from either petroleum or natural gas, and many products originally made from coal tar are now made from petrochemicals produced from the refining of oil.

Plastic America

There are other nonenergy products from oil that make more intuitive sense such as lubricants or asphalt. However, we can't talk about oil

A Few Petroleum Products

Ammonia
Anesthetics
Antifreeze
Antihistamines
Antiseptics
Artificial limbs
Artificial turf
Aspirin

Balloons
Ballpoint pens
Beach umbrellas
Boats

Candles
Car enamel
Carpets
Cassettes
Caulking
Clothing ink
Cold cream
Combs
Cortisone
Crayons
Credit cards

Dashboards
Denture adhesive
Dentures
Deodorant
Detergents
Dishwashing liquids
Dyes

Electrical tape
Enamel
Epoxy paint
Eyeglasses

Faucet washers
Fan belts
Fertilizers
Fishing lures
Fishing rods
Floor wax
Food preservatives
Footballs

Glycerin
Golf balls
Guitar strings

Hair coloring
Hair curlers
Hand lotion
Hearing aids
Heart valves
House paint

Ice chests
Ice cube trays
Insect repellent
Insecticides

Life jackets
Lipstick
LP records
Luggage

Motorcycle helmets
Movie film

Nail polish
Nylon rope

Oil filters

Pantyhose
Parachutes
Paint brushes
Paint rollers

Percolators
Perfumes
Petroleum jelly
Pillows
Plastic wood
Plywood adhesive
Putty

Refrigerants
Roofing
Rubber cement
Rubbing alcohol

Safety glass
Shampoo
Shaving cream
Shoe polish
Shoes
Shower curtains
Skis
Soft contact lenses
Solvents

Telephones
Tennis rackets
Tents
Tires
Toilet seats
Toothbrushes
Toothpaste
Transparent tape
Trash bags

Unbreakable dishes
Upholstery

Vacuum bottles
Vitamin capsules

Water pipe

without talking about plastic. Petrochemicals are used in drugs, detergents, cosmetics, food preservatives, and synthetic rubber, but the most ubiquitous product is plastic. Some plastic is made of natural gas as well, and 6 to 8 percent of total oil and gas use goes into plastic manufacture. A little over a third of that amount is required as the energy for the manufacturing process. Thirty percent of plastic becomes packaging and 15 percent becomes building materials. A number of different chemical compounds and additives are used to change the appearance, flexibility, durability, or texture of the plastic, making it appropriate for different uses.

The availability of plastic has deeply influenced American culture. Plastic packaging has made the manufacture, shipping, and selling of all products easier and more economical, allowing for long distance product safety and security. Food items, electronics, sports equipment, nuts and bolts, and even sections of double-wide homes traveling down the highway are wrapped in plastic. Sterile medical supplies such as catheters, gloves, bandages, and syringes (themselves made of plastic) come packaged in single-use plastic packages, eliminating the less reliable and inconvenient autoclave. We buy plastic packaging ourselves to wrap sandwiches, yard waste, bake sale cookies, clothes for charity, and ultimately, the trash.

A look through the kitchen, garage, bathroom, or home office will reveal the extent of the other plastic items in our lives. There are kitchen appliances, dishes, and other implements, yard work equipment, and all kinds of storage containers. All the electronics we use are made of plastic, which, for the most part, makes them possible. Without plastic, a computer case would be made of sheet metal, with metal keys like a typewriter and vacuum tubes made of metal and glass instead of circuit boards. The same would be true for televisions and radios. Football helmets would still be made of leather as would other protective gear like knee and elbow pads. Skate wheels would be wood or steel, skis would be made of wood and the boots would be leather. Badminton birds would still have real feathers. Picnic supplies, outdoor furniture, yard games, swing sets, and many other recreational items would be made of wood, metal, glass, or natural fibers, and would consequently be luxury items because they would be a lot more expensive. The main reason we can afford to have all this stuff is that it is made out of cheap, versatile plastic.

Plastic has made home building and remodeling less expensive as well, in addition to cutting down on maintenance. Vinyl siding is now standard, as are vinyl flooring and vinyl windows. There are plastic roofing materials, gutters and fencing, as well as garden decorations, outdoor light fixtures, ponds and fountains. Many of these products are made to look like the natural products they replace with surface textures and colors to resemble stone, wood, and ceramic tile. But we know they cost only a fraction of the "real thing."

Despite its versatility, convenience, and low cost, plastic presents us with a couple of dilemmas. First, it is part of the fossil fuel picture and must therefore be considered in that context. We don't know yet what impact decreasing our use of petroleum transportation fuels will have on our access to plastic. At the very least, it is likely that prices will rise. This may reduce consumer demand for less necessary plastic items, which could weaken those industries. Costs of health care could be driven up yet again from an increase in the price of plastic because of the many products they now depend upon to deliver their services. The cost of an automobile could be affected, as would the price of a new home. Going to bio-based fuel sources to replace oil will mean coming up with ways to produce all these other products from the elemental components of wood and crop residues or other organic sources, and perhaps from the byproducts of ethanol and other biofuels.

Bioplastics are not new. It is very possible that if World War II had not occurred, Henry Ford might have developed a whole new industry around plastic made of soybeans and cellulose. As it was, the war interrupted his research to develop what was called the Soybean Car, and after the war the work was not continued. This car, made with a tubular steel frame and plastic panels, weighed two-thirds what a steel car weighed, and Ford claimed it would be safer than a car with steel panels. Among other reasons for working on this idea, Ford was interested in using agricultural products in industrial processes. Biofuels and other products from plant materials are currently regarded as offering tremendous new opportunities for economic development in rural communities.

Like petroplastic, today's bioplastics are also made using processes that separate and recombine the chemical components, and their performance is comparable. Limited quantities of bioplastics are already available using such feedstocks as corn, sugar beets, sugar cane, wheat, rice, and sweet potatoes. NatureWorks, a company owned by Cargill

Fig 7. Robert Boyer and Henry Ford with the Soybean Car, built in Greenfield Village in 1930. Robert Boyer was Ford's chief chemist at the Soybean Experimental Laboratory, where they worked to find ways that farm crops could be used in the industrial economy. Reproduced, with permission, From the Collections of The Henry Ford.

Corporation, makes 300 million pounds of bioplastic a year. The market for bioplastics is more established in Europe than in the United States, but some large companies like Whole Foods and Walmart have already begun to use bioplastic for packaging. Although bioplastic requires lower temperatures in the manufacturing process, it is still energy intensive to produce. Bioplastic, like biofuel, also raises the question about the balance between what we grow for food and what we grow for fuel and other products.

The second dilemma regarding plastic concerns the environmental and health impacts of petrochemical products. Hundreds of chemicals have been developed from fossil fuels, and because they are synthetic combinations of elements, the natural world has no way of dealing with most of them. Plastic products present an environmental problem because of both their physical presence in the landscape and the chemicals that leach out of them. Most plastic does not decompose. It only breaks down into smaller and smaller particles. This means, obviously, that all the plastic we've ever thrown away is still out there. Plastic comprises about 18 percent of municipal waste, and if compostable waste were removed, it would total close to half the volume. Much plastic waste is simply floating around as pollution in the landscape, lakes, and oceans. Its unsightliness is the least of the problem. Plastic bags, six-pack holders, and Styrofoam particles are major hazards for wildlife, particularly in lakes and oceans where fish and other marine animals become entangled and drown or mistake these items for food and eat them, frequently starving to death. Garbage dumping at sea has spread plastic waste throughout the oceans of the world.

There are two basic types of plastic. Thermoset plastic cannot be melted and reused, but Thermoplastic types can. This is why different plastics cannot be mixed together for recycling. Recycling plastic is currently an idea with limited usefulness, even if it were easier to collect. The concept of recycling is to use discarded materials to remake the same products, which works well with glass or aluminum. Discarded plastic can be used to make new products, but not the same ones. Plastic water bottles can be remanufactured to make carpet, but not new water bottles. Therefore, most "recycled" plastic products do not decrease the use of virgin plastic, which is the whole point.

Toxic chemicals used in plastics, particularly additives that provide characteristics like resistance to sunlight, can leach out over time causing environmental and health problems. The plastic of greatest concern is polyvinyl chloride, or PVC, which contains high levels of chlorine and other additives. Toxins emitted during manufacture and use include dioxin, the active ingredient in Agent Orange, and vinyl chloride, a known carcinogen. It also contains an additive called diethylhexyl phthalate, a chemical recently added to California's list of known reproductive toxicants. When PVC burns it emits hydrogen chloride gas. PVC is widely used, primarily for pipes and building materials but also for toys and other consumer products, and electrical insulation. It is economical to manufacture, but there are affordable substitutes for its many product forms. PVC is being phased out as more becomes known about its toxicity, but consumer awareness will be a big factor in how fast this happens.

PVC may be the most hazardous plastic in use, but none is completely without potential toxin emissions. Bisphenol-A or BPA, a chemical used in polycarbonate plastics, is generating controversy because of studies showing it can leach out of the plastic and interfere with human hormones, particularly in fetuses and young children. Polycarbonate plastic is used for water bottles, the lining in food and beverage cans and dental sealants. It appears that we are only beginning to understand the chemical stew of plastic and its effects on our bodies and environment.

Natural Gas: Modern Miracle

Natural gas is the modern superstar of the fossil fuels. It is certainly the cleanest when burned, producing 30 percent less carbon dioxide than oil and 45 percent less than coal. The biggest issue with natural gas is that it's almost too useful. Once the national network of pipelines was in place by the 1960s and natural gas was available in most cities, its convenience began to appeal to everyone, from industrial process engineers to hotel managers to homeowners. However, the future natural gas supply is less than predictable for meeting all these demands, and the price is likely to be volatile and rising.

Its usefulness was not initially obvious. Both manufactured coal gas and natural gas were first regarded by many as mysterious and novel. Because gas is invisible, people thought burning it was like burning air. Some people were simply afraid of gas. Historically, natural gas seeps have inspired a variety of responses. The Oracle at Delphi is now thought to have been a natural gas seep that caused hallucinations to those tending it, and there are stories in many cultures of sacred "eternal flames" where natural gas was escaping to the surface and had somehow been set alight. The tenth-century Chinese were more practical, developing methods for drilling that employed a bamboo pipe. They burned the gas to distill salt from seawater.

There were efforts in the nineteenth century to use natural gas for light and heat, but these were limited because long distance pipeline technology did not exist. Solomon Dresser, an engineer from Pennsylvania, patented a leak-proof gas line coupling in 1887 which made it possible to run the first long distance gas pipeline from a rich natural gas field in Indiana 120-miles to Chicago in 1891. Dresser's coupling, which allows sections of pipe to be connected for any distance, has needed little improvement since. Despite these occasional early successes, however,

American oil drillers considered natural gas an annoying substance that would sometimes come up from the ground with the oil and would have to be burned off or "flared."

It wasn't until about 1910 that the oil industry started taking natural gas seriously enough to become the "oil and gas" industry. Pipelines were built and drilling began in earnest. By the late 1940s natural gas was poised to replace coal and oil as the residential heating fuel of choice, and to transform the nation's industrial sector by providing both a process fuel and a raw material for products like fertilizer and chemicals. Natural gas can be found in oil fields and coal beds, but 75 percent of gas production in the United States comes from fields where it exists on its own. Natural gas must be processed to remove a variety of other substances including butane, propane, and helium, which are removed and sold as separate products. Other byproducts might include ethane, pentanes, sulphur, and nitrogen, and many other chemical substances may be present depending on the characteristics of the deposit. It is easy to understand why natural gas provides the foundation for the chemical industry.

Natural gas is colorless and odorless and is not poisonous. Because it is lighter than air it tends to dissipate rapidly into the atmosphere but it can become a safety hazard. It can asphyxiate anyone present if it builds up in an enclosed space and replaces enough oxygen. Or, if it comprises between 5 and 15 percent of the space, it will explode if ignited. This is why one of the last steps in processing natural gas is the addition of mercaptan, a smelly chemical whose unpleasant odor alerts us to gas leaks. Once the gas is refined, it is piped to tanks or old gas wells for storage until it is needed, when it is distributed through the pipeline system to customers.

So Many Uses

Because it burns cleaner than other fossil fuels and is easily transported, natural gas has become like the popular dinner guest who is invited to every party. Gas is now the primary heating fuel for homes in the United States, replacing both coal and oil. The residential sector uses about 22 percent of the nation's gas production. We also use natural gas for cooking, clothes drying, and water heating. Currently, over 60 percent of homes use natural gas including 70 percent of new homes.

Natural gas has also become essential for a number of energy-intensive industries, primarily the chemical industry, which produces the raw feed stocks for our heavily synthetic culture. Over 40 percent of the natural gas consumed in this country goes directly or indirectly to making paper, asphalt, plastics, paints and solvents, cement, fertilizers, and to refine petroleum, and it is also used widely for food processing and other industrial heating uses. Businesses and institutions use about 15 percent for heating, cooking, laundry, and heating water for everything from swimming pools to car washes. Gas has long been considered cheaper than electricity for heating, and even with the volatility of gas prices over the years this has generally been true.

Much of the recent growth in natural gas use has occurred because of a supply blip in the eighties and nineties. In response to the 1979 oil crisis, a rash of exploration for oil found a lot of natural gas instead. Prices fell and expectations rose, particularly in the electric utility industry, which was looking for a clean source of fuel for generating electricity in order to meet the stringent requirements of the Clean Air Act. Electricity generation has become the fastest growing segment of natural gas consumption, and today almost 15 percent of the gas supply is generating electricity. Growth in this sector will likely continue because utilities have discovered that it is easier and faster to site a natural gas generation plant than a coal plant.

In the past, natural gas exploration and production have followed a pattern based on consumption trends. Because most gas has been used in the winter when heating needs increase, the supply could be caught up in the summer when demand is less. However, with new uses altering this annual supply-demand cycle, the industry has been struggling to keep up. For example, with more electricity being generated with natural gas, the seasonal air conditioning load is placing new pressure on the summer gas supply.

The transportation industry is now looking to natural gas to help address the issue of U.S. dependence on foreign oil for vehicle fuel. Ninety-nine percent of the U.S. natural gas supply currently comes from North America with 85 percent from within U.S. borders and the rest primarily coming from Canada. Greater dependence on natural gas to produce transportation fuel will drastically increase overall demand year round. There is already a fleet of close to 150,000 vehicles on American roads that burn compressed natural gas (CNG), primarily buses, taxis, and other commercial vehicles, and gas trade associations and other groups are advocating for more. The aggressive new ethanol industry wants natural gas for the energy-intensive process that makes corn into biofuels for America's gas tanks. Furthermore, the fertilizer needed to boost the productivity of growing corn is made with, and by, natural gas. And those who dream of a "hydrogen economy" usually fail to mention that the primary source of commercial hydrogen today is from a process of steam reforming natural gas.

So Uncertain the Supply

Natural gas is highly attractive to energy interests in the United States because it is both domestic and clean. However, there are already flashing signs that the U.S. natural gas supply is overbooked. Onshore domestic natural gas reserves peaked in the 1980s, and since the late 1990s the price of natural gas has increased over 500 percent. This is because we are using it faster than we are extracting it. To add to the difficulties, Canada may soon decide to cap its exports of natural gas to the United States in order to protect its reserves for the needs of Canadian citizens who are particularly dependent on it for heat in the winter months.

It seems that, just as we have decided natural gas is the fuel of the future, or at the very least the transition fuel toward the future, we find we are facing a supply crisis. According to the American Chemistry Council, the U.S. chemical industry, which is dependent on large quantities of natural gas, has lost a number of manufacturing plants and thousands of jobs because of high natural gas prices. Domestic production is down from other related industries such as paper and fertilizer, with companies either shutting their doors or moving overseas where natural gas prices are lower. The fertilizer industry is a case in point. The productivity of American agriculture is legendary, and its methods have been exported extensively abroad. Much of this productivity rests on the ubiquitous use of synthetic nitrogen fertilizer, available only since the early twentieth century.

Nitrogen is abundant in the atmosphere but not in a form that plants can use. Some crops, like soybeans and other legumes, have bacteria on their roots that convert atmospheric nitrogen to a plant-friendly form. On the other hand, corn is greedy for synthetic nitrogen fertilizer and wouldn't grow so lush and green without it. The nitrogen fertilizers in use today, including ammonium nitrate, ammonium sulfate, and urea, are all made from anhydrous ammonia. Anhydrous ammonia is also used in plastics manufacturing and as a refrigerant, and to promote growth of beneficial industrial bacteria. This substance is the outcome of the Claude-Haber ammonia synthesis process by which natural gas is combined under high temperature and pressure with the atmosphere. Natural gas provides both hydrogen to the mix and heat to the process, and it makes up 90 percent of the cost of fertilizer production. Sharply increasing gas prices have contributed to a decline in the American fertilizer industry as lower gas prices in other countries have made U.S. fertilizer less competitive. Some farmers have switched crops based on rising fertilizer costs.

Up until recently, our growing dependence on natural gas for so many things has not raised much concern. On the contrary, because it is the cleanest fossil fuel and because we get most of our supply from North America, it has been viewed as the fall-back fuel. However, when hurricanes Katrina and Rita tore through the Gulf Coast in 2005 disrupting so much of the oil and gas infrastructure and contributing to distressingly high price spikes, the fragility of the supply became evident and high-volume users began looking for solutions.

The American Chemistry Council, which represents the industry that uses more than 10 percent of the U.S. natural gas supply, would like to see the U.S. Congress lift its moratorium on natural gas drilling on the outer continental shelf, a ban that has been in effect since 1981. They are concerned that natural gas prices in the United States have become the highest in the world. With natural gas being essential to producing chemicals, and chemicals being necessary for producing over 95 percent of all products in the economy, the United States, which has already become a net importer of chemicals, could take a permanent back seat in the world economy. The Sierra Club, recognizing the potential necessity of lifting the ban but wary of the free-for-all that might result, is advocating for a careful planning process that would protect more fragile areas of the marine environment and set standards to keep drilling impacts to a minimum. Without knowing yet what actions will be taken, we should keep in mind that while geologists are optimistic about finding new fields on the outer shelf it is not presently known exactly how much is out there or how long it would last.

Liquified Natural Gas (LNG)

A second solution, importing liquified natural gas from overseas, addresses a whole different set of challenges. Currently 1 percent of the U.S. natural gas supply is imported as LNG and that quantity is expected to rise to 3 percent by 2010. However, depending on imported LNG from elsewhere will require a vast and expensive infrastructure that does not yet, and may never, exist for a number of complex reasons.

Natural gas is a regional fuel, meaning it can be transported as a gas easily over land through pipelines, but with greater difficulty and expense across oceans. Even overland lines of great length, like the one that would be required to tap the natural gas fields on Alaska's North Slope, are expensive and hard to engineer. The most practical way to move natural gas across an ocean is to liquefy it and ship it in tankers. Natural gas liquefies at minus 260 degrees Fahrenheit and, when liquid, occupies one-six-hundredth of the space it needs in its gaseous form. This means that ships, trucks, and storage facilities must be kept at that temperature to preserve the natural gas in its liquid state. These ships are high-tech, refrigerated, double-hulled wonders designed for maximum safety, but they sport a high price tag. Capital investment adds up, with the liquefaction plant at the port of origin and the regasification plants in U.S. ports where the gas is offloaded to the pipeline. Current exporters of LNG include Algeria, Indonesia, Malaysia, Qatar, and Trinidad, with Australia, Iran, Nigeria, and Russia as potential major exporters.

The importing of LNG has become a contentious issue in the United States. A number of regasification facilities currently exist, but many more would be needed if we expect to depend on this source. The ideal port for building such a plant would be well sheltered, deep, and broad, allowing room to maneuver the large LNG tankers. Most U.S. ports that fit this description are already heavily developed and populated. Even though the Federal Energy Regulatory Commission (FERC) oversees a complex siting process for these plants that considers safety and security procedures, it is impossible to predict the potential level of risk from terrorism.

In addition to the capital expense of building a new energy infrastructure and the security risks this would entail, an essential problem remains. We would need to import LNG from some of the same politically volatile nations where we are currently getting petroleum, thus creating a new dilemma—dependence on foreign natural gas, and we will be in competition with other nations also seeking to import it. While there remains a lot of natural gas in the world, getting at it and transporting it will only become more expensive and difficult. The long-term solution can only be making the transition away from natural gas, starting with those steps most easily taken. These would be making a deep commitment to energy efficiency in all sectors and to significant development of renewably generated electricity. Sorting out our use of biomass resources for fuels, plastics, chemicals, and other products will be more complex but will be essential as a long-term solution.

Uranium and Nuclear Power

Uranium is not a fossil fuel, but it isn't a renewable energy resource either. Because it has more in common with coal than it has with wind or solar energy, uranium, used to generate nuclear power, is included here. Like coal, uranium must be mined, processed, and transported, all requiring fossil energy resources. Similarly, coal and uranium each present significant environmental challenges, although with coal it's primarily the total CO_2 output, while with uranium it's the toxic levels of radiation, from mining to the disposal of nuclear waste. The third reason to include uranium with fossil fuels is the scale of its generating plants. Nuclear power plants are typically large-scale, centralized facilities like coal plants, requiring a complex planning and siting process and an extended construction schedule.

The first use of nuclear energy was the U.S. bombing of Hiroshima and Nagasaki in August 1945, with the goal of ending the war with Japan. From one bomb dropped on each city, fatalities are estimated at around 140,000 in Hiroshima and 74,000 in Nagasaki—almost all civilians. Both cities were completely destroyed. Nuclear energy was a fearsome new weapon that could cause an insidious style of destruction and death with far fewer actual armaments. Shortly after the war, the United States went on to develop and test the hydrogen bomb, an even more powerful weapon. The stage was set for the Cold War and the race to nuclear superiority. Needless to say, the world had become a fearful place with those nations in possession of nuclear technology seen as holding all the cards. Small nations were feeling particularly vulnerable.

President Eisenhower recognized the international anxiety and realized that the United States needed to take responsibility for unleashing nuclear energy into the world. He responded with a remarkable plan to share nuclear energy among nations. Eisenhower gave his famous address, "Atoms for Peace," to the United Nations General Assembly in 1953, urging the nations of the world to work together to use nuclear energy for positive purposes, and to create an International Atomic Energy Agency to oversee its development. He then worked to create the Atoms for Peace program, which provided federal funding to develop peacetime uses for nuclear energy, primarily in medicine and agriculture and to generate electric power. The program also provided nuclear reactors to nations who signed an agreement to use them for peaceful purposes only. The ultimate success of his program is still being debated, but it did lead to development of the Nuclear Non-proliferation Treaty in 1968 and the Comprehensive Test Ban Treaty in 1996.

It's quite possible that we started generating electricity with nuclear power only because we felt a compelling need to do something peaceful with this source of energy that had such otherwise horrifying potential, not because it was a particularly efficient way to generate electricity. If we had not invented nuclear bombs, we might not have embraced nuclear electricity with such enthusiastic optimism. At the time, promoters were claiming that nuclear power was "too cheap to meter," but we knew very little then about the effects of radiation and had not foreseen the challenge of storing the radioactive waste.

Riding on a wave of peaceful neutrons and hefty federal subsidies, the nuclear power industry grew fast in the 1950s and 1960s, despite nagging questions about safety and radioactive waste. There are currently 436 nuclear power plants in the world, with 103 of these in the United States. The U.S. plants generate about 20 percent of the nation's electricity. France has 59 plants and Japan, 55. After the partial meltdown at Three Mile Island in 1979, no new plants were ordered in the United States as the public became wary of the technology. A similar reaction occurred in Europe after the Chernobyl explosions in Russia in 1986. More recently, the potential for nuclear weapons development by terrorist organizations or hostile governments has been raising significant concern. As these safety, security, and waste issues have emerged, the price tag for nuclear power has risen dramatically and it is no longer considered as competitive as it once was. However, the climate change issue and the need to reduce CO_2 emissions have given nuclear energy advocates new hope. There is growing interest in nuclear power once again, primarily because the generation process does not emit CO_2. Not all agree, however, that this valuable and timely attribute can trump its other challenges. Of these, environmental impacts and hazards to health and safety are the greatest.

Generating Electricity with Uranium

Uranium deposits contain both radioactive U235 and U238. U235 is the most amenable to fission, a molecular chain reaction that releases heat. Natural uranium deposits are less than one percent U235, so the ore must first be refined into uranium oxide, called yellowcake, and then enriched to a level of 3 to 5 percent to be used in a commercial nuclear reactor. Highly enriched yellowcake is 20 percent pure, and bomb-grade yellowcake must be 90 percent pure. The 3-percent yellowcake is made into pellets that are implanted into long metal rods. These fuel rods are lowered into the water in the reactor core. As fission occurs, the heat that is generated is used to make steam, which powers a steam turbine to generate electricity. A nuclear power plant is much like a coal-fired plant in this regard—they both heat water to make steam for a steam turbine.

Nuclear Waste

A typical nuclear reactor produces between twenty and thirty tons of high-level nuclear waste each year. High-level waste is mostly spent fuel rods, but it also includes equipment used in the generation process located in the reactor core. Everything close to the nuclear chain reaction becomes imbedded with radioactivity. Fission of uranium creates new isotopes and elements including cesium, strontium, and plutonium. The latter is the radioactive element used for making bombs and will remain radioactive for 240,000 years, or 12,000 generations. By the year 2000, nuclear reactors had created 201,000 tons of high-level nuclear waste. No permanent, long-term storage facility has yet been built to store this waste, and it remains in temporary storage on sites where the reactors are located. Construction and operation of nuclear plants produces other toxic waste as well, including acids, solvents, heavy metals, fluorides, mercury, and asbestos.

The U.S. government intends to create a storage facility for high-level nuclear waste at Yucca Mountain, a ridge located within the boundaries of the Nevada Test Site, one hundred miles northwest of Las Vegas. This site was first identified as a potential storage location in 1978, and the U.S. Department of Energy has been seeking approval since then. The most recent projected date for opening the repository is 2017, if all permits are approved. However, the Yucca Mountain facility will not be enough. The commercial nuclear power plants currently in operation will have produced enough high-level waste by 2014 to fill it to the capacity it is initially permitted to store.

There has been much study of the Yucca Mountain site to ascertain its suitability as a long-term storage facility. Many engineers and geologists agree that it's a good choice, but political winds have blown back and forth on whether Yucca Mountain will open. It has stirred up a controversy emblematic of the whole nuclear energy issue, particularly in the State of Nevada. Emotions run high among Nevadans who lived through the nuclear testing era when bombs were exploded in their back yards, and 87 percent of the state's land belongs to the federal government. An overwhelming majority of Nevada residents oppose the repository, resenting the idea that they will be forced to host the nation's nuclear waste. Native American tribes in the area are objecting to the dumping of hazardous waste in the landscape they claim as their cultural heritage. Not only Nevadans but the rest of the country as well are very uncomfortable with the idea that trucks and trains will be rumbling through their towns carrying casks of waste to Nevada, regardless of assurances from the U.S. DOE that the containers are safe. A big part of the problem is that people recognize how seriously powerful nuclear energy is, and they are unsure about trusting its regulation, security, and storage to an essentially political institution with a constantly shifting agenda. Who do we trust with something that will be deadly for 240,000 years?

Hazardous to Health and Communities

The harmful radiation from a nuclear reaction is emissions in the form of alpha, beta, and gamma rays, each stronger and more hazardous than the last. These rays are all capable of destroying living cells. Alpha rays will not penetrate healthy skin but are harmful if ingested or inhaled. Gamma rays require protection on the order of a concrete wall to avoid penetration. Health effects depend on the amount and duration of the radiation exposure. High-level exposure over a few minutes, like that of Hiroshima survivors, causes acute radiation syndrome (ARS). This starts with nausea, diarrhea, bone marrow depletion, and flu-like symptoms and can lead to death in weeks or months. Low-level exposure over long periods of time has a cumulative effect on the body and the impacts may not become evident for many years. This kind of exposure can lead to cancer, cataracts, and decreased fertility.

The low-level health hazards of uranium typically occur around the mining and milling of the ore. Uranium is mined from both surface and underground mines. While uranium occurs in many places, the largest known deposits are in northern Saskatchewan in Canada, and there are also substantial deposits in Australia. Tailings from these and other mines leach radioactivity into the soil and ground water and can contaminate local vegetation and structures in the landscape. In a number of locations, local residents have suffered ill effects from mining operations, even when the mining takes place in relatively isolated areas. In Niger and Namibia, tailings dumped in the desert contaminated water used by local nomadic tribes. Uranium mining contaminated fisheries essential to local communities in northern Canada.

During the nuclear power boom in the United States after World War II, substantial deposits of uranium were discovered on the Navajo Reservation in the area called the Colorado Plateau, which includes corners of New Mexico, Colorado, Utah, and Arizona. The Navajo tribe was initially glad to have the local jobs, but the miners were not informed of the hazards of close contact with the uranium ore. There were few safety precautions offered to the workers, and eventually high rates of cancer, pulmonary fibrosis, pneumoconiosis, silicosis, tuberculosis, birth defects, and kidney damage became evident among miners and their families. There were also accidental releases of mill wastes and ongoing seepage from tailings ponds that contaminated both surface and ground water in the area. Wells that tap these contaminated aquifers are used for drinking and irrigating crops. The desert wind blows radioactive dust from piles of tailings at abandoned mines over great distances. The piles also produce radon gas, a known carcinogen that continues to affect local communities.

Uranium mines were closing down by the seventies, but by that time about eleven hundred small uranium mines had been dug on the Navajo

Reservation. Most were simply abandoned, with tunnels left open and tailing piles unprotected. Fewer than five hundred mine sites have been cleaned up, but local communities are still located near abandoned mine sites, using the land for grazing and other subsistence farming. The Radiation Exposure Compensation Act of 1990 was passed to compensate people who had been exposed to nuclear radiation as the result of testing or mining. There were some cross-cultural paperwork difficulties for the Navajo Nation that resulted in many affected families being unable to establish their claims to the satisfaction of the federal government.

Because of this past experience, many members of the Navajo Nation are justifiably leery of further uranium mining on the reservation. In 2005 the Tribal Council voted to ban mining when a company called Hydro Resources, Inc. submitted a new license application to the Nuclear Regulatory Commission. However, some tribal members are in favor of reopening mining operations because of the jobs that would become available. It is not unusual for such a controversy to emerge in poor communities that sit on rich deposits of extractive resources. It is often difficult for these communities to contend with the economic and political pressure from outside without incurring internal social damage. Despite their official commitment to the ban, the Navajo Nation may not have the last word on whether this mining company moves forward, particularly if demand for uranium continues to rise. When there has been a lot of money at stake in the past, the interests of the extractive energy industries have frequently overridden the interests of local communities.

Safety and Security

Since the first nuclear power station was built we have endured the risk of an accident that would cause a radioactivity-enhanced disaster. So far, we have experienced the partial meltdown at Three Mile Island and the explosions at Chernobyl, and there have been a number of reports of other close calls. We are told that newer reactors are designed to be safer, with added back-up containment systems and advanced controls. Surely we have learned a lot since the early days about making things safe. Unfortunately, the world in general has become less safe in the meantime. The reduced hazards in modern plant design have now been replaced by the increased threats from terrorism, and potentially from extreme weather events such as hurricanes, tornadoes, and floods. Developing new layers of protection from these dangers simply adds to the already high price tag on nuclear power plants.

An even greater security question centers on the issue of nuclear proliferation. Plutonium, or Pu-239, is used to make nuclear bombs. It happens to be a waste product of fission and can be separated from the spent uranium and used for doing precisely that. Reprocessing spent fuel is also a way of reducing nuclear waste, because the uranium can then be used again in the reactor. However, it is this reprocessing step that creates the weak security link between civilian and military applications of nuclear energy. Nations that have the capacity to reprocess the spent nuclear fuels from their nuclear power plants are also able to produce their own supply of weapons-grade plutonium, or at the very least, a quantity of plutonium that must then be secured and tracked.

Advanced reactor designs called closed fuel cycle systems separate the plutonium from the unused uranium, allowing its reuse. These systems are in place elsewhere in the world, but until recently the United States has allowed only "once-through" thermal reactor technology. The Bush administration's Global Nuclear Energy Partnership (GNEP) has embraced the idea of reprocessing to help relieve the waste issue. However, many experts feel the advantages of more efficient use of the fuel do not outweigh the significant security risks posed by the additional production of plutonium. In its position paper *Nuclear Power and Global Warming*, the Union of Concerned Scientists states that the GNEP, in its focus on promoting closed fuel cycle reprocessing technologies, "shows no prospect of creating a proliferation-resistant nuclear fuel cycle and is encouraging other nations to engage in dangerous plutonium fuel operations."[5] The report recommends that Congress restore the previous policy banning reprocessing. An interdisciplinary study from the Massachusetts Institute of Technology titled *The Future of Nuclear Power* draws similar conclusions about the security risks of reprocessing systems.[6]

While the MIT study gives greater weight to the possible role of nuclear power in reducing carbon emissions, both these reports find nuclear energy problematic as a solution to future electricity demands. In addition to the safety, security, and waste issues, the high cost of building nuclear power plants may make it difficult to find investors when there are other clean energy sources to develop. Nuclear energy may be pricing itself out of the market. Also, the time required to design, site, and build a plant will mean that we won't have enough plants to make a dent in overall carbon emissions for twenty years or more.

Nuclear energy is not simply one of many choices for generating electricity. Its potential for destruction is immense, whether through human error, aggression, or natural disaster. This fact places it in a different category from all other energy sources, and no reasonable, technical argument can completely diminish its menace. The costs of nuclear energy go beyond economic or environmental considerations because the controversy that erupts over siting waste dumps or generating plants extracts a social toll in the form of divided communities and perceived and actual injustices that may never heal. If we begin to account for the social costs of promoting a technology that many do not trust, we may decide to invest our effort in other generation choices.

NOTES

1. U.S. EPA, Carbon Sequestration in Agriculture and Forestry, http://www.epa.gov/sequestration/index.html, downloaded December 31, 2007.

2. The Environmental Literacy Council, *The Horse in the Urban Environment*, http://www.enviroliteracy.org/article.php/578.html, downloaded July 8, 2007.

3. Union of Concerned Scientists, *Backgrounder: How Oil Works*, www.ucsusa.org/clean_energy/fossil_fuels/offmen-how-oil-works.html, downloaded July 2, 2007.

4. The Pew Research Center for the People and the Press, *News Release: Economic Pessimism Grows, Gas Prices Pinch*. Washington, D.C: The Pew Research Center, September 2005. http://people-press.org/reports/display.php3?PageID=998, downloaded July 10, 2007.

5. Union of Concerned Scientists, *Nuclear Power and Global Warming*, March 2007. Downloaded on July 18, 2007 from http://www.ucsusa.org/assets/documents/global_warming/nnp.pdf.

6. Massachusetts Institute of Technology, *The Future of Nuclear Power*, 2003. Downloaded on July 18, 2007 from http://web.mit.edu/nuclearpower/pdf/nuclearpower-summary.pdf.

RESOURCES

Coal

The American Coal Foundation (ACF)
The ACF is an educational nonprofit organization located in Washington, D.C., that produces coal-related materials and programs for teachers and students.
http://www.teachcoal.org/

Natural Gas

The American Gas Foundation (AGA)
The AGA is a nonprofit organization that funds research about energy and environmental public policy related to natural gas.
http://www.gasfoundation.org/

Oil

The American Petroleum Institute (API)
The API is a national trade association whose members represent all facets of the U.S. oil and gas industry.
http://www.api.org

Nuclear Energy

The American Nuclear Society (ANS)
American Nuclear Society is a non-profit scientific and educational organization which promotes awareness and understanding of the applications of nuclear science and technology.
http://www.ans.org

Fossil Energy Resources From the U.S. Government

The U.S. Department of Energy (US DOE) fossil energy web page:
http://www.fossil.energy.gov

The U.S. Energy Information Administration (US EIA) Energy Kid's Page on Non-Renewable Energy
http://www.eia.doe.gov/kids/energyfacts/sources/non-renewable/nonrenewable.html

The National Energy Technology Laboratory (NETL)
With sites in Morgantown, West Virginia; Pittsburgh, Pennsylvania; Tulsa, Oklahoma; Albany, Oregon; and Fairbanks, Alaska, NETL is the U.S. Department of Energy's national laboratory dedicated to fossil fuel technologies.
http://www.netl.doe.gov

For Further Information About Peak Oil and Global Warming

Natural Resources Defense Council (NRDC)
An independent nonprofit environmental action organization that offers information about a variety of energy issues and impacts.
http://www.nrdc.org

Pew Center on Global Climate Change
An independent nonprofit founded in 1998 to provide credible information and innovative solutions to the global warming crisis.
http://www.pewclimate.org

Union of Concerned Scientists (UCS)
An independent nonprofit, this organization is known as a reliable source of independent scientific analysis of environmental and energy issues.
http://www.ucsusa.org

CHAPTER TWO:

The Electricity Grid

In May 1941, the American folk music legend Woody Guthrie accepted a commission from the U.S. Department of the Interior to write a series of songs in praise of the huge new Bonneville Dam. This was the first federal hydroelectric project, built on the Columbia River in the state of Washington. By June he had produced twenty-six songs and was paid $266.

That a passionate champion of the common people like Woody Guthrie would so fully embrace a huge new technology is a sign of how important Americans considered electricity in the first half of the twentieth century. Electricity meant lights, washing machines, irons and toasters, along with another device almost as powerful as light—the radio. Suddenly, news was instantaneous and universal. It's no wonder that Woody Guthrie lent his support to national electrification. It was the key to empowerment for a democratic people. Electricity brought us into the light of the modern age, relieving untold drudgery and serving as both a social leveler and symbol of progress.

Electricity in Our Lives

The greatest gift of the fossil fuel age is electricity. Let's just pause to realize what a difference this miracle has made in our lives. Forget for the moment how we generate and sell it, or what challenges we face in its future production. We will discuss those aspects presently. First, let us simply be grateful for electricity.

In the present time, we can bless the ingenuity of the inventors who have brought us light bulbs, microwave ovens, computers, and medical instrument sterilizers. Electricity makes possible hot food, cold drinks, and television. It controls our heating and cooling systems, refrigerates our medicines, sews our clothing and freezes our food. Without it our emergency rooms would be useless, we would need a traffic cop at every street corner to direct traffic, and there would be no such things as

Fig 8. The spillway of the Bonneville Dam under construction. The Bonneville Dam is located on the Columbia River between Oregon and Washington. Construction began in 1933 and took the U.S. Corps of Engineers almost ten years to complete. Besides generating power, it is used for flood control and navigation. Photo courtesy of the Bonneville Power Administration

fast food or Monday night football. With electric power we can dry our hair, saw lumber, pump water out of the ground, and make ice all summer long.

Only when we go camping in the wilds or endure a power outage do we experience what it might have been like before. After the sun goes down, the only sources of light would have been candles, kerosene lamps, or the light from the fireplace. Doing the wash would mean pumping the water by hand and carrying buckets to the wash tubs or the stove. Clothes would be washed, rinsed, and wrung out by hand or crank wringer. Ironing them would be done with heavy "irons" heated on the stove or in the fire. Cooking would require chopping or carrying wood or coal for the stove, and all the water used in the kitchen would once again be pumped and carried by hand. Keeping clean would be a weekly bath with more hand-pumped and carried water, and this didn't get thrown out until every family member had used it.

These were only the household chores. Factories, sawmills, and granaries were located beside rivers because then they could use the mechanical power from water wheels, but most things were made by hand. There were no motors to lift heavy machinery or to run lathes, potters' wheels, printing presses, or other tools. Human muscle powered these operations. Work on the farm was similarly muscle-power intensive. Some work, like plowing and clearing of land, was done with the help of horses, mules, or oxen. But feeding and watering the livestock was once again manual labor.

In the evolving urban industrial society of the early twentieth century, household help was not typical in middle-class homes and the average woman did all her own housework. The coal or wood used for cooking and heating would smoke when it burned, leaving a dirty film over everything. Kerosene lamps produced fumes and were often the cause of accidental fires. Electricity not only made many new conveniences available, it eliminated a lot of the grime and the danger. Life really was noticeably easier, and definitely more modern. Electricity was seen as a miracle that relieved drudgery in both urban and rural America and brought opportunities for a better life.

Electricity also made city streets safer as electric street lighting began to replace gas lights in the 1880s. The first functional electric trolley system was built in Baltimore in 1885. The first electric elevator was installed in New York City in 1889 by the Otis Company, forever changing city skylines everywhere. Electric power for industry made it possible to run factories round the clock. All kinds of new production machinery became available, and factories were no longer dependent on the availability of water or steam power nearby. Without electricity, Henry Ford's assembly line production would not have happened, and automobiles would have remained the custom built toys of rich hobbyists.

Electricity birthed the consumer age, beginning in urban areas. Electrification of America's cities came about through sophisticated advertising campaigns promoting electricity's cleanliness, healthiness, and its ability to relieve the burdens of housework. Electricity became the symbol of success and prosperity, allowing Americans to embrace the new and seductive concept of "leisure time." Most cities and towns were electrified between the 1880s and the 1930s. Half of rural homes were electrified by 1945, and over 90 percent had been connected by 1953.

War production during the 1940s boosted electricity usage to a new high. By the time the 1950s began, America was ready to build its postwar prosperity, and electricity was there to power it, from industries to homes. The new sophisticated image of the American woman was the 1950s housewife in her modern tract home with all the electric appliances at her command. During the war, women had to work, replacing men in the factories while they went off to fight. When their husbands returned to civilian life, women were pushed back into their traditional roles in the home, but the consolation prize was the convenience of all the fashionable, new electric appliances. The miracle of electricity has only been near universally available in the United States for the last seventy-five years. The era before electricity is still within living memory. Yet we

Fig 9. "Leisure with Electricity" advertisement from 1946. When World War II ended, electric utilities promoted their product to a nation of women returning to homemaking from jobs in the war industries. Electric appliances contributed a great deal to the American social structure and lifestyle of the fifties. Courtesy of Library of Congress, Prints & Photographs Division, Theodor Horydczak Collection, [reproduction number LC-H814-T-1999-047 DLC]

now have come to take its necessity and availability completely for granted. We don't realize how much electricity has even become a matter of life and death. In the aftermath of Hurricane Katrina in 2005, loss of electric power hampered rescue efforts, communication, and security and made conditions unbearable for the survivors who were stuck in the flooded city of New Orleans. Electricity doesn't just make life easier, in today's society, it makes living possible.

Some History of the Industry

Electricity was first employed in the United States for street lighting. Small systems sprang up in varying locations during the early 1880s, including San Francisco, Cleveland, Baltimore, Philadelphia, Kansas City; Portland, Maine; and Des Moines, Iowa. Entrepreneurs, using a variety of system designs, started small companies offering wiring and lights for a price in a limited area near their "dynamo" or generating station. It was during this time that the first publicly owned municipal utilities were founded to provide street lighting in downtown areas. Most of these early systems employed arc lights, a lighting type different from the incandescent bulbs of today. Arc lights were eventually replaced by incandescent lighting, but arc light technology provided the basis for today's mercury vapor, sodium vapor, fluorescent, and metal halide lamps.

At this point there was no consistency in equipment nor was there organization among proponents or developers. In short, it was the usual chaos present in a promising and rapidly developing new technological opportunity. There wasn't really an electricity industry as such, but it was obvious that something big was on the horizon. The time was ripe for a visionary to pull it all together.

Thomas Edison, Electricity Baron

It was not his many inventions, but rather his vision for an electricity industry, that would turn out to be Thomas Edison's greatest contribution to life as we know it. Edison is known as an inventor, but he was primarily an ambitious and brilliant entrepreneur. His work with electricity convinced him of its potential, particularly for providing light. But he recognized that the electric systems already in use were not competitive with gas on a large, replicable scale. He studied the successful and established gas industry, which was currently providing most urban lighting, to see what he could adapt to an electric production and distribution system. Edison also envisioned driving the gas light industry out of business by replacing it with his own electricity empire. He saw powering electric motors as an additional use, making the whole system more efficient because lights would be needed at night, but motors would operate during the day when manufacturing firms were open for business. Edison systematically went about acquiring or inventing all the components he would need to establish a cost-effective, central electric power system, including switching gear, lamp holders, and control devices. Edison didn't actually invent the incandescent light bulb, but he bought and improved upon the design patented in 1875 by Canadian inventors Henry Woodward and Mathew Evans.

Edison built his first commercial central station in 1882 on Pearl Street in Manhattan. The coal-fired dynamo (a reciprocating engine generator) at Pearl Street Station could produce 100 kilowatts of power that operated 1,200 light bulbs, and at a price competitive with gas lighting. Today, a typical coal-fired base load power plant produces 500 megawatts, 5,000 times larger than the Pearl Street Station generator. Not surprisingly, the most complicated part of Edison's project was convincing the City of New York to allow installation of the cables and wiring to deliver the power. Light bulbs were expensive, the neighbors complained about the coal smoke and fumes, and Edison had to figure out how to meter the electricity each customer used. The project did not turn a profit for two years. It had, however, illuminated a new dream. Edison founded the Edison Company for Isolated Lighting in 1883 to build electric power stations across the country.

Before the modern electricity industry would truly be born, however, there were a number of turning points that could be characterized as technical or economic, but in reality they hinged on the actions of a few powerful personalities. First was the adoption of alternating current (AC) as the standard. Edison's Pearl Street Station delivered direct current (DC), which worked well for 110-volt electricity to power light bulbs but not for the higher voltage applications industry would began to demand. Furthermore, direct current could not be economically moved over great distances because it couldn't be transformed to higher voltages that made this possible. A fierce competition broke out between Edison and George Westinghouse, another inventor, and founder of Westinghouse Electric. Westinghouse promoted the work of his resident genius, Nikola Tesla, who had developed alternating current and was convinced it would be more effective and efficient for powering a larger network. Despite some very mean-spirited campaigning by Edison, who felt threatened by Westinghouse's competition, AC power gradually became the standard. Incidentally, Nikola Tesla was a man greatly ahead of his time in a number of his ideas about energy. He recognized the non-renewable nature of coal and other fossil fuels and promoted solar, hydropower, and wind energy.

Edison insisted that coal supplies were virtually endless, an opinion the coal barons were more than willing to share. This entrepreneurial determination overpowered Tesla's futuristic thinking, and he would eventually be considered a brilliant but eccentric and impractical inventor, while Edison became the Father of Electricity. Edison also managed to shout down one of his primary investors, J. P. Morgan, regarding the structure his electricity empire would ultimately assume. Their disagreement centered on how they would make money from electricity. Morgan wanted to sell equipment for generating electricity while Edison wanted to own the equipment itself and sell the electricity. By 1902, the total amount of electric power generated in the United States was evenly divided between central stations that served customers in the surrounding area, and on-site generation plants installed by individual customers, mostly powering motorized equipment in factories. Edison's monopolistic vision received an unexpected boost from the demands of World War I. By the end of the war in 1918 and after the intensive increase in demand for electricity to produce goods for war, most manufacturers found it more cost-effective to purchase their electricity from central stations.

Edison wanted to control the electric power industry, and to this end he worked hard to convince America that it needed electricity. Edison himself didn't invent modern marketing, but the burgeoning electricity industry employed it effectively. Industry entrepreneurs knew they wouldn't make money on this highly capital-intensive venture until they had captured the whole of America's imagination. There were two primary ways this was done. First they promoted electricity as clean, healthy, and capable of relieving women's housework drudgery. Second, they made electricity as cheap as possible so that no one could resist it.

Technological developments improved system efficiency and lowered costs dramatically during the first three decades of the twentieth century. Key to this was the invention of the steam turbine with its giant economies of scale. Steam turbines continue to be used in today's coal-fired and nuclear generating plants. Improvements in organizational efficiency, however, contributed to greater cost effectiveness as well. This was brought about by another visionary who understood the magnitude of electricity's potential and the need to extend cost efficiency beyond technology to the way the whole industry operated.

Insull's Influential Market Model

Samuel Insull, a protégé of Thomas Edison, rose to control Chicago Edison, a huge Midwestern power company. He was an influential leader in the industry, which had attracted big money players. Competition was fierce and sometimes not so gentlemanly. Insull saw this capitalistic anarchy as detrimental to the interests of the industry as a whole. After some persuasive arguments to fellow power moguls, he eventually convinced them to voluntarily submit to emerging state regulation as a means of eliminating expensive duplication of effort and fights over territory. He also argued that the industry would gain respectability under the watchful eye of the government, and private companies, as franchisees of the State, would be able to shoulder out both new start-up companies and the small but growing municipal utility movement that was gaining popular support. His main point was that if they agreed to stand together on an even playing field, they could all focus on attracting more customers to defray their high capital investment in equipment and wiring. Insull then advocated continually raising efficiency and lowering prices to make electricity available to as many people as possible and too cheap to refuse. He knew the industry needed a huge customer base to make money. This strategy was a stroke of genius, and growth in the industry exploded.

Insull and his contemporaries were not simply selling electricity, they were selling cheap electricity, a marketing strategy entirely appropriate to building the customer base for a new industry. Today, both utilities and their regulators have continued to emphasize low rates as the highest priority despite the fact that almost everyone everywhere has already been connected to the grid. Customers still expect cheap electricity, and now expect a totally reliable and abundant supply. Unfortunately, this outdated market model has made itself unsustainable by ignoring what we have come to know about the environmental impacts of fossil fuel generators and avoiding the additional price tag.

The Industry: Where Are We Today?

The delivery of a reliable and sufficient supply of electric power has become a profound responsibility. Because electricity can't be economically stored on a large scale, the system must generate electricity at roughly the same rate that it is being used, and on a continual basis, taking into account both periodic and unexpected demand surges. This might be the normal increase in demand from commercial buildings during business hours or residential air conditioner use late on summer afternoons. A sudden heat wave, however, would require that additional supplies be available at a moment's notice. Utility planners must juggle

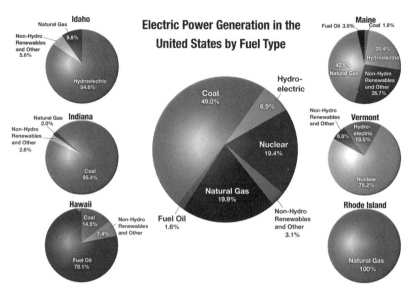

Fig 10. *Electric Power Generation in the United States by Fuel Type, 2006. While coal supplies about half the fuel nationally for generating electricity, individual states vary widely in their fuel mix, as illustrated by the six examples shown. In the chart, "Non-Hydro Renewables and Other" includes agricultural waste, geothermal, landfill gas, municipal solid waste, solar, wind, wood, and fossil fuel byproducts.* Source of Data: U.S. Department of Energy, Energy Information Administration, 2006

generation from base load plants, adding or subtracting peaking plants as necessary, and continually watch out for other contingencies like downed lines from bad weather or equipment failure at substations.

Planners are also juggling sources of power and the costs to produce it when it's needed. Most utilities don't own all the generation capacity they need, but have contracts with independent power producers (IPPs) and other utilities. They must pay more for the power from a natural gas-fired peaking plant on a hot July afternoon than they do for electricity from the base load coal-fired plant at night when demand is low. When utilities talk about whether their capacity is adequate, they refer to having enough generating capacity available to keep everything running when demand is at its highest, for whatever reason. Their forecasting is purposely conservative in order to maintain a reserve. These forecasts usually include the assumption that one or more unexpected problems will happen every day. Planners also include regular maintenance for generating plants in their forecasts.

Utility planning, and the technology that supports it, has reached an extraordinary level of sophistication, all in the interests of providing the level of reliability upon which we have come to depend. It's not surprising that the industry tends to regard its projections as self-fulfilling prophecies, and indeed, so far that has been a fairly clear path for them to follow. With a lead time of five to ten years to site and build a new coal-fired power plant, they see it as necessary to invest considerable effort into long-range projections. Investor-owned utilities in particular have adopted the perspective that they have no control over how much electricity gets used or will be demanded in the future. All they can do is respond to that demand, and second-guess what people will need in coming years, through the use of ever more complex and sophisticated planning models and computer programs. This approach has worked well for the last forty years or so, conveniently supporting the bottom line as well, but the world is changing fast.

Makeup of the Industry

The electric utility industry in the United States is huge. According to U.S. Energy Information Administration (EIA) data from 2004, the industry serves over 136,000,000 customers. Utilities sell wholesale and retail electricity to the residential, commercial, and industrial markets. They frequently sell natural gas and other energy services. Many publicly owned utilities also offer water and sewage treatment, as well as internet and other communications services.

In the United States today, there are three primary utility business structures that serve retail electricity customers, along with several types of wholesale power producers and transmission companies. Retail utilities include publicly owned utilities, investor-owned utilities, and electric cooperatives. Together these companies form an interconnected generation, transmission, and distribution network that provides the nation with highly reliable electric power.

The federal government utilities operate about 180 power plants, which are under the authority of three primary agencies: the Army Corps of Engineers, the Bureau of Reclamation, and the Tennessee Valley Authority. Most of these projects are large hydroelectric plants built originally in the first half of the twentieth century to serve irrigation and flood control functions as well. Federal power plants were developed to provide

Electric Utilities in the United States

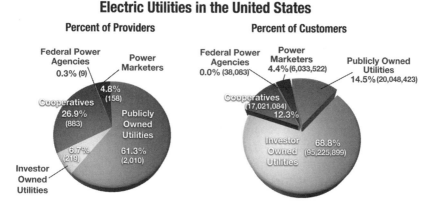

Percent of Providers
Percent of Customers

Fig. 11. Electric Utilities it the Unites States, 2007 The source for the data in this table is the *2007–2008 Annual Directory and Statistical Report of the American Public Power Association.* Reproduced by permission.

power to municipal utilities, rural electric coops, and other nonprofit utilities and to sell power at rates that merely recover costs of production. These agencies have other bulk power customers as well. For example, the Tennessee Valley Authority sells power to 53 large industrial customers, 6 federal facilities, and 12 neighboring investor-owned utilities in addition to its 108 municipal and 50 cooperative utility customers.

Publicly Owned Utilities

Publicly owned utilities are actually local or state government agencies that operate independently as nonprofit corporations. They are usually municipal utilities, but they can also be organized as State public power districts, irrigation districts, or other state organizations. These utilities are governed by an elected board of commissioners, and as part of the local government infrastructure they are subject to the political scrutiny of the voters who form the customer base. They are community owned, formed to provide essential services to local residents, and are usually able to do so at a lower cost because of their nonprofit status. According to the American Public Power Association (APPA), the national organization of publicly owned utilities, their core values are "community service, customer orientation and local control." Some of the currently operating municipal utilities were among the first utilities established over 125 years ago.

There is considerable pride in the local ownership and control of the electricity supply, and a recognition that keeping up with energy developments, trends, and customer demands is essential to their future.

Investor-Owned Utilities

An investor-owned utility (IOU) is a for-profit corporation that is granted a franchise for a specific geographical service territory by the state regulatory agency, but it must agree to serve all customers in that area. The level and types of service, and their rates, are closely regulated by the state where the service territory lies. IOUs can serve customers in more than one state, but they must answer to all the appropriate state regulators. Most IOUs currently provide electricity generation, transmission, and distribution, although this is gradually changing with the restructuring of the industry. For example, some states' regulators are requiring utilities to "unbundle" the three basic services with the goal of keeping transmission and distribution as regulated while allowing generation to operate in an open market.

For the investor-owned utilities, the emphasis on load projections is in line with their function as for-profit corporations and their focus on paying dividends to their shareholders. It is not in their financial interest to get involved in cutting electricity consumption or experimenting with new generation technologies when they can rely on the far cheaper, established fossil fuel technologies. Investor-owned utilities sell electricity to almost 70 percent of the nation. Consequently, their focus on building and operating huge, central, coal-fired generation plants has colored the view most people have of what is possible when it comes to producing or buying electric energy.

Rural Electric Cooperatives

In the cooperative electric utilities, customers are also members, and the tradition of active member participation has been part of the Rural Electric Cooperative structure since its beginnings. The development of the Rural Electric Cooperatives is an amazing tale of rural citizens bootstrapping their way into electrification by way of Rural Electrification Administration (REA) loans and expertise, and a lot of local determination, cooperation, and elbow grease. This history of member ownership

America's Electric Cooperative Network

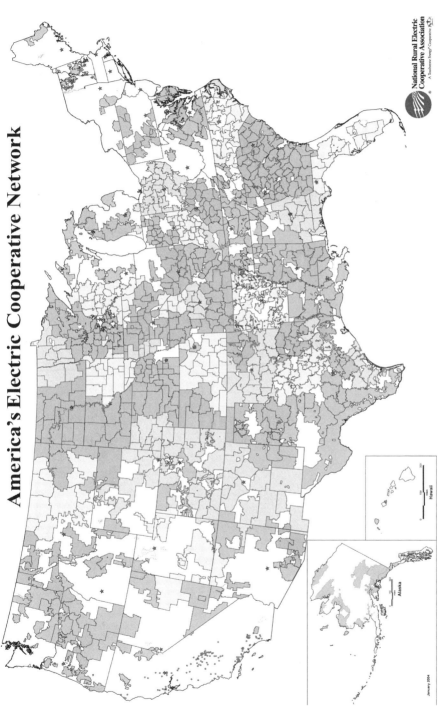

National Rural Electric
Cooperative Association
A Touchstone Energy® Cooperative

January 2004

Hawaii

Alaska

Fig. 12. America's Electric Cooperative Network. This map illustrates the extensive geographic areas served by the nation's 883 Rural Electric Cooperatives. Map courtesy of the National Rural Electric Cooperative Association

and pride continues today in coops across the country. The story of rural electrification is worth mentioning here because it provides a valuable precedent for the concept of sustainable energy as a local enterprise as well as offering proof that the federal government can actually help people to help themselves.

In the opening decades of the twentieth century, the availability of electricity in urban areas created a sharp divide between city dwellers and rural Americans in their standard of living and social position. In a major societal shift, rural towns, previously seen as the backbone of Jeffersonian democracy and the foundation of American society, fell behind as electrification passed them by. By 1932, more than 80 percent of urban homes were electrified, but only about 10 percent of America's rural residents had electricity, and these were the wealthy farmers or those near towns. Farming not only continued to mean a grueling life of primarily manual labor, it became identified with backwardness and poverty. Even though the disastrous droughts in the 1920s and the Great Depression of the 1930s perhaps dealt more serious blows to the viability of farm life at the time, the lack of electric power contributed mightily to degrading the image of the farm family and its place in the American social order. Anyone who could do so fled to the "bright lights" of the city.

The main reason for the disparity in electric service was that the distance between customers in the country was so much greater then that between urban customers. Commercial utilities determined that it was not cost effective to run electric lines to reach isolated farmsteads, and they had no interest in serving that market. They were not even interested in the low-interest loans offered by the REA. Anxious to get things going, the REA offered loans instead to organized groups of farmers and rural residents who were already accustomed to cooperative business structures for marketing crops and sharing equipment. Response was tremendous. In 1935, when the REA was first signed into existence by President Roosevelt, only about 10 percent of farms had power. In 1953, less than twenty years later, over 90 percent were connected. Much of the work, from organization to installation to operation, was done by the cooperative members themselves. To keep costs down, the REA worked closely with the coops to develop an assembly line style of installation and to implement volume purchasing of materials. The REA loan program is still in operation as the Rural Utilities Service, under the U.S. Department of Agriculture. Rural electric cooperatives currently serve

40 million people in forty-seven states. While only a few electric cooperatives have chosen to be on the cutting edge of clean energy so far, they are certainly in the position to respond to their members' desire to move in that direction. Many coops interpret their mission of responding to member needs as contributing to the sustainability of their communities. Electric coops in various parts of the country are beginning to take a leading role in development of renewable energy, particularly wind, and biomass from both farm and forest. Their membership structure gives them flexibility for responding directly to customer priorities, and to utilize local resources.

Because publicly owned utilities and cooperatives are nonprofit, member, or voter driven organizations, they play a very different role with those they serve than the investor-owned utilities, who must answer to their investors first. However, even though there are more nonprofit than for-profit utilities, the percentage of customers served by the nonprofits is considerably smaller and less influential than the investor-owned utilities in the overall picture. Even so, both the cooperatives and the municipals are healthy, well-established, and seamless segments of the national grid. It is interesting that the electricity industry has both nonprofit and for-profit components operating side by side, illustrating the difficulty of defining the industry's essential purpose. There are those who feel that the electric utility industry is simply a sizeable segment of the national industrial economy that should be making quarterly profits and basing its decisions on traditional economic factors. Others feel it is a service for providing a commodity necessary to public health and safety for which economics is only one of several long-term planning strategies.

Since the early days of utility entrepreneurship, electricity has gone from being something of a novelty for the well-to-do to becoming a national necessity. Access to electricity is as much a determinant of a minimum standard of living as food, shelter, and clean water. From a societal perspective, the systems that protect our health and secure our streets from both criminal activity and anarchy in traffic could not function without it.

The utilities do an excellent job of supplying the nation with power. They have done well with meeting power demands and anticipating future rates of growth by constructing new plants and new transmission lines. They work hard to maintain "four nines," the utility jargon for 99.9999 percent reliability. The electricity industry in the United States is highly organized and regulated to provide utility service that is the envy of many other countries. But the predominant voice of the electric utility industry in this country is a private sector voice that answers to the short-term priorities of its stockholders rather than to the communities it serves. Its long-term decisions are beginning to differ from community energy concerns, particularly those regarding the environment.

Regulation and Deregulation

In some countries, the electric industry is owned and run by the national government. In the United States, the majority of the industry is in private hands, but these utilities are heavily regulated by the government. Although electric utilities are subject to regulation by local, state, and federal agencies, it is the state agencies that primarily regulate the investor-owned utilities. A few states have also established jurisdiction over publicly owned utilities and cooperatives. Every state is different. Many state regulatory agencies oversee natural gas utilities and communications industry operations as well. Large investor-owned utilities with service territories in more than one state must comply with the rules of each state with regard to their customers in that state. Regulation has created a structure that represents the public interest with regard to availability and affordability of electric power. Over time it has evolved into a complex and powerful means to assure the financial soundness of the investor-owned utilities, as well as assuring they meet certain obligations for what has become a public institution every bit as necessary as hospitals, police departments, and fire stations.

By 1916, thirty-three states had established regulatory agencies, primarily in response to the rapid spread of utility territories beyond state boundaries, as well as concerns about consistent rates and service among the private utilities. The establishment of regulatory agencies was official recognition of the "natural monopoly" the emerging electric grid and its captive customers represented. Theoretically, regulation was expected to substitute for the benign attributes of healthy competition by establishing reasonable rates and quality service for customers. Electric service had become both a virtual monopoly and a commodity necessary to health and safety. Private utility companies applied for franchises to provide a defined level of service to customers in a particular territory, with rates and financial structure dictated by the regulators.

Table 1
Government Regulatory Roles in the Electric Industry

Local Government Elected Municipal Utility Commissions	State Government Regulatory Commissions	Federal Government Federal Energy Regulatory Commission (FERC)
Jurisdiction over municipal utilities	Approval of plant & transmission line construction	Wholesale electric rates
	Retail rate levels	Qualifying facility inter- connection (PURPA)
	Jurisdiction over investor owned utilities (IOUs)	Licensing of Hydropower facilities
	Municipal & cooperative utilities (some states only)	Nuclear safety & waste disposal
		Environmental regulations
		Natural gas pricing, oil pipeline rates, and gas pipeline certification

The Federal Role

The federal government has stepped in from time to time to regulate issues of national scope as well as those that go beyond control by individual states. The first major pieces of federal regulatory legislation were the Public Utility Holding Act of 1935, which addressed financial abuses by utilities that were creating large, questionable holding companies, and the Federal Power Act, also enacted in 1935 to regulate interstate transmission and sale of power. In 1936, President Roosevelt signed the Rural Electrification Act, which created the Rural Electrification Administration (REA), whose purpose was to lend money and technical assistance to organizations that would bring electricity to the critically underserved rural areas of the nation.

Then, in 1954, the federal government responded to the quest for peaceful uses of nuclear power with the Atomic Energy Act, which brought about the potential for commercial development of nuclear power generation. The first nuclear plant was up and running by 1957 in

Shippingport, Pennsylvania. In 1969, the National Environmental Policy Act was the first official recognition of sulfur dioxide (SO_2) as a serious pollutant. This law required that new plant permit applications include environmental impact statements (EIS). Coal-fired power plants fell under further scrutiny with the Clean Air Act of 1970 and subsequent amendments in 1977 requiring reductions of SO_2 and other emissions. The issues of water quality and utility waste disposal (for both hazardous and nonhazardous materials) followed, with the Federal Water Pollution Control Act of 1972 and the Resource Conservation and Recovery Act of 1976. This run of environmentally related legislation concluded in 1978 when President Carter signed the National Energy Act, which included the National Energy Conservation Policy Act statute requiring utilities to offer information to their customers about reducing electricity use through conservation. Another statute under the National Energy Act was the Public Utility Regulatory Policies Act, known as PURPA, designed to promote cleaner, more efficient use of fossil fuels and to develop more renewable sources of electric generation.

Deregulation

State regulatory agencies and investor-owned utilities have been unsettled for the last fifteen years or so because of federal government pressure to deregulate the industry. National efforts at utility restructuring essentially began with PURPA. On the heels of the oil embargo of 1974, the federal government wanted to open up the "natural monopoly" of the utility industry in order to increase efficiency of fossil fuel generation and development of renewable energy resources. The strategy was to allow a broader range of generating facilities to "qualify" for interconnection to the grid, such as industrial cogeneration units and small-scale renewable energy systems. According to PURPA, if small facilities met federal standards and had excess power to sell, the utilities were obliged to buy it. In other words, a utility could not refuse to allow a manufacturing facility with its own cogeneration unit that was generating excess electricity, or a wind turbine producing more power than the site needed, to sell the extra power to the utility grid. The utilities were not pleased with this mandate because it meant spending time and money dealing with small systems that would not pay back the interconnection investment. Big or small, every interconnection has to meet minimal safety requirements.

As a result of creative foot dragging on the part of many utilities, PURPA did not immediately expand the cogeneration or renewable energy industries at the time. Nor were state utility commissions placing a high priority on net metering or interconnection procedures for small renewable energy systems, because demand was very low. Most of the early customers for residential solar or wind systems were located in remote areas too far from power lines to be connected. However, PURPA did lay a solid foundation for the net metering laws, regulated interconnection guidelines, and renewable energy portfolio standards and tariffs that states would late develop. Under a net metering law, utilities must credit customers who generate surplus electricity (and feed it back to the grid) at the same rate they pay to buy utility electricity, but the law usually stipulates a maximum size for eligible systems. Interconnection guidelines assure consistency, fairness, and timeliness of the process for customers of all utilities within the state who wish to interconnect a generation system to the grid. Renewable portfolio standards (RPS) require utilities to have in their portfolio of generation sources a certain minimum of power generated by renewable energy, usually expressed as a percentage of the total. Renewable energy tariffs are rates that reflect higher perceived value for green energy, a roundabout way of recognizing the environmental impacts of fossil fuels and making renewable energy more economical to generate. These state-level regulatory tools are all essential to establishing the very different strategy of integrating renewable energy into the utility generation mix in both rural and urban areas.

Despite the fact the PURPA will be thirty years old in 2008, state commissions are only now beginning to recognize the value that small renewable systems can offer the utility grid. The Network for New Energy Choices, a nonprofit organization promoting clean energy options, has published the second edition of their 2006 report, *Freeing the Grid, 2007 Edition.*[1] In this report they grade the states for their net metering and interconnection policies, using for comparison the models developed by the Interstate Renewable Energy Council. These models were designed to assist state commissions in removing the regulatory barriers to mainstream adoption of renewable generation systems. Of the thirty-nine states (including Washington, D.C.) that have net metering laws, twelve received a grade of A or B and thirteen earned a dismal D or F. Grades for state interconnection rules were lower. No state received an A for its rule, only two states out of thirty-five received a B, and twenty-one scored D or F. The Network for New Energy Choices does comment that their 2007 report reflects considerable progress among states since they published the 2006 edition of *Freeing the Grid,* and they project this progress will continue as demand for green power grows.

Another piece of federal legislation has been critical to moving the industry toward deregulation. The Energy Policy Act of 1992, known as EPACT, expands the categories of eligible power producers to include nonutility wholesalers and broadens the authority of the Federal Energy Regulatory Commission to introduce greater cordiality into "wheeling," the utility term for moving electricity over transmission lines belonging to other utilities. In other words, utilities must be more cooperative about letting other utilities transmit electricity across their lines at reasonable cost.

State utility commissions are under pressure to find ways in which the electric utility industry can learn to thrive in a more competitive business environment while continuing to provide what is essentially a public service. Some states are experimenting with requirements for the unbundling of services and establishing separate transmission utilities that would remain under regulation. The federal government wants states to hammer out the details that would make it possible for customers to choose the supplier they prefer and still have electricity delivered to them over local distribution lines. No state can yet claim to have established a completely successful system to accomplish this goal, although several have tried.

The IOUs had close to sixty years of mostly profitable prosperity, taking place in a predictable and cooperative regulatory environment. With the new federal legislation in place there has been far less certainty about where profits will come from and what their future relationships with state regulatory agencies will be. IOUs are not accustomed to planning the kind of business development strategies other nonregulated industries pursue, and state regulators are similarly unfamiliar with thinking of electricity as a product in a competitive market. Deregulation could be called a work in progress, because a number of thorny issues have emerged that are slow to resolve. With utilities competing to sell power and not tied to an agreement to provide even-handed rates and service, the labor-intensive residential power market tends to experience higher and higher rates, while the high-volume industrial customers enjoy the greatest price breaks. Utilities also find that capital investment in new plants and transmission lines puts them at a competitive disadvantage, so these projects get postponed, leading to a weakened infrastructure.

The Energy of the Future

More and more communities are setting high expectations for environmental quality and greater local control of energy sources. Despite a number of forward-thinking private utilities that recognize opportunities in the changing utility picture, their investor-oriented structure and sheer size do not offer the flexibility that environmental leadership will require. The traditional publicly owned and cooperative utilities offer successful models for local energy production and distribution that may address emerging community energy priorities. They seem to be preparing for it. The APPA 2002 Task Force Report "Public Power in the 21st Century" concludes with the statement, "No matter what your utility's successes to date, the old ways of doing business will not be sufficient for the future."[2] But publicly owned utilities serve only about 17 percent of electric customers nationwide. Clearly, utility customers themselves, whether residential, commercial, or industrial, as well as utility stockholders, municipalities, and government agencies, will need to bring their voices to the conversation.

Most investor-owned utilities continue to operate in the capital-intensive, centralized, fossil fuel infrastructure that built the industry from its early days. But making money the old-fashioned way will begin to earn diminishing returns as fossil fuels become more expensive. Yes, the United States still has considerable coal reserves, but this coal must be moved over vast distances by diesel-powered trains from mine to generating plant. It would be far more efficient to locate generation plants closer to where power is needed, just as many other products and services will need to be produced locally in a sustainable economy.

The most pressing environmental sustainability goals, such as lowering greenhouse gas emissions and cutting other forms of pollution like mercury and particulates, are fossil fuel related. Coal-fired generation of electricity currently produces almost 40 percent of America's CO_2 emissions. However, the pricing of electricity continues to focus on keeping it cheap for the consumer rather than on charging a price that fully reflects the costs of generating it. The U.S. Environmental Protection

Agency has been slow to designate CO_2 as a pollutant to be regulated. Once it does, coal-fired generation will become considerably more expensive as environmental regulations limit emissions of carbon dioxide (CO_2) along with other regulated pollutants.

The coal industry is promoting what it calls "clean coal technologies" for use in generating plants. These technologies process the coal through washing or gasifying and then burn it in a multistage process that controls the sulphur dioxide. They are being designed to capture and sequester or store CO_2 emissions, some by injecting the CO_2 deep into the ground. There are no clean coal systems up and running yet that actually sequester carbon, and it is unclear how long it will be until clean coal is ready for commercial prime time. Many other questions have also been raised about this heavily promoted idea, such as how reliable CO_2 storage would be or how enforcement of compliance with regulations would be managed.

The primary advantage of clean coal technology is that it would allow the current centralized generation infrastructure to remain in place and business to carry on as usual. Naturally, coal and utility investors are interested in maintaining this status quo. It's true, as they say, that if all proposed coal-fired generation plants were built as clean coal facilities, air quality would improve and carbon emissions would go down. However, as the primary solution to meeting our future demand for electricity, it's an expensive and risky path. Knowing what we know now about the negative aspects of burning coal and the rising costs of transporting it, we may find that it's more worth our time to pursue other energy resources for our electricity and to explore the new economic opportunities offered by a more diverse and renewable energy portfolio.

Attracting new customers is no longer a major objective for utilities. On the contrary, they are working hard to keep up with the rising demand for electricity. With that in mind, we need to consider whether we can continue to afford cheap electricity. Right now the cheapest electricity comes from the oldest, dirtiest coal plants, many of which were scheduled to retire but are kept on line because they were grandfathered in before pollution control devices were required. Meanwhile, the rate of children's asthma is skyrocketing, fresh water fish are increasingly inedible, and the climate is changing. Because electric rates are kept low by ignoring environmental "externalities," we tend not to value the energy, and we use it pretty thoughtlessly. On the other hand, many people are willing to pay more for clean electricity and for the chance to make that statement. Green power programs, which sell renewably generated electricity for a slightly higher price, are gaining popularity all over the country. It's even starting to pay off. During a recent natural gas price spike, customers of a Colorado utility, who were paying a premium to buy wind power, paid less for their electricity than neighbors who were charged the going electric rate along with the additional surcharge for the natural gas.

Future Possibilities

Electricity promises to be the medium of energy for the future, for both transportation and nonmobile uses. It is clean at the outlet and endlessly useful. We are developing ways to generate it without fossil fuels, and energy efficiency can show us that it's not how much we use that counts, it's what we do with it. However, sustainable electric power will mean more than efficiency or renewable sources of generation. The electric power industry will be undergoing changes as well, potentially moving from a centralized production model to one more local and distributed. Long-distance hauling of coal in diesel-powered trains will become less and less cost effective as the price of oil rises. Natural gas supplies will become less reliable. Stiffer environmental regulations loom as the effects of greenhouse gasses, fine particulate pollution and mercury are more fully recognized. Perhaps most compelling for the utilities themselves is the movement toward electric utility deregulation initiated by the federal government in the 1970s. Opening the industry up to market forces is a complex and contentious idea, but in the long run it may accelerate sustainable energy goals.

Distributed Generation

Distributed generation brings electricity customers into the generation mix by connecting many small, on-site systems to the grid. Their gradual adoption will alter the centralized infrastructure and make electricity generation a more local enterprise. Customers can either sell their surplus and draw what they need from the grid, or they can generate their own electricity when it's cheaper than buying it from the utility.

J. P. Morgan would be happy to see all the technologies available today that make individual ownership of electric generation not only possible but rapidly becoming practical and cost effective. In his day, owning a coal-fired "dynamo" for generating the electricity needed to light a home or business really was a messy and labor-intensive prospect. No wonder Edison's service model for utility development seemed like the better idea. Not so now with solar PV panels that have no moving parts, produce no smoke or other pollution, and can be expected to continue operating with minimal maintenance for thirty years or more. Or, there are the highly efficient, natural gas–fired microturbines that can provide peaking or backup power on site for hospitals or manufacturing facilities. Then there is the methane from landfills or sewage treatment plants that municipalities can use to generate electricity for their operations. These are all examples of "distributed generation" technologies located on the electricity user's property that, if interconnected with the utility grid, become part of a system of distributed generation.

The concept of distributed generation looks good for a number of reasons and consequently enjoys a broad base of support. First, it is a structure that is gradually emerging from the federal PURPA legislation of 1978 by interconnecting small power producers (SPPs), which must use at least 75 percent renewable resources for generation. It has fallen to the individual states to determine exactly how this would be accomplished within their borders, but they share similar issues with regard to encouraging renewable energy systems: consistent interconnection agreements, net metering, renewable energy portfolio standards, and setting rates that recognize the extra value of clean energy by paying a higher price per kilowatt hour.

Energy Security

Whether planning for natural disasters or potential terrorism, the security of the electricity supply is of major concern. Distributed generation can be incorporated into this type of planning because it responds to security concerns by spreading the generation sources around. An energy security plan could make use of the concept of "microgrids," areas where a number of homes or businesses have their own generation systems and could be hooked together in their own distribution grid but interconnected to the utility network at one point. Interconnection of small systems occurs at distribution lines rather than transmission lines. Distribution systems will need to be redesigned to accommodate large numbers of small generators coming on line. Distributed generation can take pressure off the high voltage transmission line network by providing more power locally where it's needed. Imagine a new housing development that generates a portion or all of its own power, making it unnecessary for the local transmission line to be upgraded.

Finally, power that is generated locally keeps energy dollars within the local economy. Large numbers of small systems increase installation and maintenance jobs within the community. There is more potential for job creation in the renewable energy industry than in the fossil fuel or centralized generating plant industries where the goal over the years has been to replace labor with machinery and automated controls.

What's Next?

The future of the utility industry is by no means predictable. However, with utility deregulation, environmental constraints, and volatility in fossil fuel markets, it will need to be more capable of responding to local needs and resources. There will be numerous opportunities for communities to get involved in generating and distributing much of their own electricity as well as in taking greater responsibility for using it wisely. Municipal energy efficiency standards, requirements for developers of both residential and commercial projects to generate part of their own electricity, or community investment in utility-scale wind turbine projects are a few examples. New economic models are emerging as well such as "community aggregation," a strategy that has been instituted in some states as part of deregulation where a municipality can purchase bulk power on behalf of its individual residents, thereby competing with industrial customers for cheaper rates. It's still possible to form a publicly owned utility or a rural electric cooperative if a sufficient number of local residents wish to do so. Communities wishing to pursue greater energy self-reliance will have many tools to work with as the industry is restructured.

Generation will change, but we will still need the grid. The utility grid will continue to provide the structure for quality and reliability, but will be retooled to focus more on expanding local distribution functions. The "natural monopoly" first extolled by the early entrepreneurs can be redefined to mean a structure that recognizes the power of diverse ownership and cooperation to make the most efficient use of local energy resources.

NOTES

1. The Network for New Energy Choices, *Freeing the Grid: 2007 Edition.* Available for Online Download: www.newenergychoices.org/uploads/FreeingTheGrid2007_report .pdf, downloaded December 21, 2007.

2. American Public Power Association, *2002 Task Force Report: Public Power in the 21st Century, 2002,* http://www.appanet.org/aboutpublic/index.cfm?ItemNumber=9869&sn .ItemNumber=12477, downloaded October 5, 2007.

RESOURCES

Nonprofit and Industry Websites

American Public Power Association (APPA)
A nonprofit service organization established in 1940 whose members are over 2000 community-owned electric utilities across the country. APPA promotes the policies of publicly owned utilities and provides services to its members.
http://www.appanet.org

Edison Electric Institute (EEI)
EEI is the nonprofit member organization for investor-owned utilities. Founded in 1933, it represents its members in advocating for public policy, and provides members with re-search, business services, and other support.
http://www.eei.org

National Association of Regulatory Utility Commissioners (NARUC)
A nonprofit organization founded in 1889, NARUC's members are the governmental agencies that regulate telecommunications and electric and water utilities across the country. NARUC's mission is to serve in the public interest to improve delivery of util-ity services. Their home page includes a link to state commission web pages.
http://www.naruc.org/index.cfm

National Rural Electric Cooperative Association (NRECA)
NRECA is the national service organization for over 900 rural electric cooperatives in 47 states. It was founded in 1942 to address wartime electric issues in rural communities and has been advocating for and assisting cooperatives ever since.
http://www.nreca.org

Network for New Energy Choices (NNEC)
NNEC is an independent nonprofit organization that promotes safe, clean, and environmentally responsible energy options, advocating energy conservation, energy efficiency, and renewable energy.
http://www.newenergychoices.org

U.S. Government Websites

Energy Information Administration (EIA)
Electricity: U.S. Data, Reports, Analyses, and Forecasts.
http://www.eia.doe.gov/fuelelectric.html

Bonneville Power Administration (BPA)
The BPA in Portland, Oregon, is a federal agency that serves the northwest region, offering transmission services and wholesale, primarily hydroelectric, power, and working to preserve the environmental quality of the region.
http://www.bpa.gov/corporate

Tennessee Valley Authority (TVA)
TVA, another federal agency, is the largest public power provider in the country. Founded in 1933 by the TVA Act, this agency has overseen many issues having an impact on its constituency in the Tennessee Valley, including power production, navigation, flood control, malaria prevention, reforestation, and erosion control.
http://www.tva.gov

CHAPTER THREE:

Efficiency and Conservation
THE SCIENCE AND ART OF USING LESS

As the second half of the twentieth century began, the American economy was poised for greatness. We'd survived the crippling Great Depression of the 1930s, and by 1945 we had successfully vanquished fascism in Europe and the militancy of the Japanese emperor. In the process we built a vast public and private energy infrastructure that was all set to jumpstart the unprecedented prosperity of the 1950s. After the deprivations of depression and war, Americans were ready to fall in love with modern conveniences like all electric kitchens and television, and, of course, great big cars. For twenty-five years we prospered unchecked and reveled in our seemingly endless, and conveniently available, energy resources.

It was the oil embargo of the 1970s that rudely awakened us to the wasteful energy habits we had acquired. For the first time in our history we lost total control of our energy supply, and we faced potential difficulties in getting the energy we needed at the price we had come to expect. Both the public and private sectors responded with efforts to use less energy while maintaining economic prosperity. These efforts have met with varying degrees of success, but all have been important in helping us understand what works. What began as a desperate energy reduction movement in the 1970s has been simmering quietly in the background ever since, gathering data, experience, and wisdom. The energy efficiency and conservation movement introduced the idea that how we use energy is as critical as where we get it. Now, thirty years later, we face the necessity of dramatically reducing energy-based carbon emissions for the sake of sustaining a livable environment. The efforts begun in the 1970s provide us with a firm foundation for understanding how we can meet our steep new energy reduction goals.

Two Approaches to Using Less

There are two different ways to reduce energy use: "energy efficiency" and "energy conservation." Efficiency means getting more work from the same energy source, as in using a 25-watt compact fluorescent light to produce the light level of a 100-watt incandescent bulb. Conservation means choosing to use less energy, as in remembering to turn off the compact fluorescent light when leaving the room. Understanding the difference between efficiency and conservation is important because we have gotten quite good at the first but have made very little headway with the second. This is easy to understand because becoming more efficient is a quantifiable, technical challenge that leans heavily on science while improving the bottom line. On the other hand, conservation implies creating a new style of living by employing conscious intent and imagination to reduce use of resources while maintaining one's quality of life, which is much less measurable.

In the past thirty years, pursuing the science of efficiency has been far easier than promoting the art of conservation. As a result, regulations, codes, standards, and programs have emerged from both the public and private sectors that have markedly improved the nation's energy efficiency. This work represents a good head start on energy sustainability for the future and therefore merits a closer look.

Making Lemonade in the Seventies

The Arab oil-producing nations handed us a lemon by declaring the oil embargo in 1973, and the federal government responded with its own recipe for lemonade by starting an era of effective energy policy leadership that would last through several presidential administrations. Starting in the seventies and extending into the eighties, Washington funded energy efficiency research and alternative energy technology development and created public energy education programs. The federal government also began issuing a series of appliance, vehicle, and equipment efficiency standards that are still paying dividends. Private sector and state government efforts supported the federal policy during this period of energy consciousness-raising as well, which resulted in utilities creating demand-side management programs for their customers, development of a number of local and national building codes and standards, and the beginnings of what would become the green building movement we know today. As a result of leadership in Washington, the concepts of energy efficiency and conservation permanently entered the public mind and vocabulary.

The oil embargo itself was a temporary shock, but the real surprise was the long-term response of the national economy to this federal policy response. The low key but persistent work by both private and public sector energy activists began to alter the national energy consumption pattern. Forecasters had predicted that energy use would increase about 3.6 percent every year as it had during the previous two decades, moving somewhat in parallel with growth in the Gross Domestic Product (GDP). Indeed, economists had always assumed that energy consumption marched in lock step with economic growth. However, the unexpected increase in energy efficiency firmly decoupled GDP from energy use as shown in the table from the Pew Center on Global Climate Change below. The U.S. Department of Energy forecasters had predicted a level of energy use almost double what it turned out to be. Thanks to the steady plodding behind the scenes by state energy offices, federal agencies, manufacturers, utilities, and nonprofit advocacy organizations to educate the public and establish new efficiency standards, the relationship between energy use and economic growth was permanently changed.

An Era of Technological Efficiency

Despite the public's lukewarm response to programs promoting energy frugality in the seventies once the long lines for gas had disappeared, government agencies at both state and federal levels began quietly instituting a variety of efficiency standards. As a result, the greatest progress in reducing energy use over the last thirty years has actually been through the design of more efficient products. Whether it's household appliances, automobiles, or industrial processes, energy productivity has taken a quantum leap since the seventies. This is not to say that we've achieved maximum technological efficiency, but the concept is now firmly rooted in the public mind. With a well-insulated home, an ENERGY STAR–qualified furnace and compact fluorescent lights, one can be prudent with energy and still live quite comfortably. Improved technology has allowed us to reduce our energy use with little or no effort.

We've welcomed the energy-saving appliance standards, codes, and utility programs that have emerged over the years. But our acceptance or adoption of energy-saving measures in our own lives has not been driven by the kind of urgency President Carter was encouraging us to feel when he spoke of "the moral equivalent of war." Perhaps we like the

Historic Growth in U.S.
GDP and Energy Consumption | 1949 - 1999

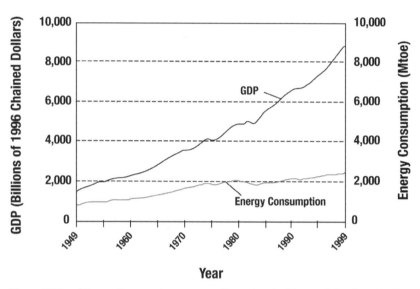

Fig. 13. *GDP and Energy Consumption, 1949–1999* Reproduced with permission from the Pew Center on Global Climate Change

idea of saving something on our utility bills, but for the most part, we have not felt there was any other compelling reason to pay attention to our energy use.

The Seventies Correction

There are many reasons why our relationship with energy changed dramatically in the seventies. Because of generally low energy prices, individual consumption had crept up to reach its highest level ever. The percentage of oil imported from elsewhere began to climb steadily as domestic reserves peaked. Electric utilities had raised generation efficiency to the point of diminishing returns, but they were faced with rising costs and a rapidly growing demand for power. At the same time, domestic energy policy was low on the nation's priority list. As they say on Wall Street, we were due for a correction.

The Moral Equivalent of War

Tonight I want to have an unpleasant talk with you about a problem unprecedented in our history. With the exception of preventing war, this is the greatest challenge our country will face during our lifetimes. The energy crisis has not yet overwhelmed us, but it will if we do not act quickly.

We simply must balance our demand for energy with our rapidly shrinking resources. By acting now, we can control our future instead of letting the future control us.

Our decision about energy will test the character of the American people and the ability of the President and the Congress to govern. This difficult effort will be the "moral equivalent of war"—except that we will be uniting our efforts to build and not destroy.

—Excerpts from the televised speech delivered by
President Jimmy Carter to the nation on April 18, 1977

When Carter made his speech in 1977, a subliminal shock went through the American psyche. It was the first time a peacetime president spoke of making sacrifices for the sake of the future, and it was the first time an American political leader recognized global energy resources as finite, or the need for America to think about curbing its energy appetite. Although he addressed only the imminent scarcity of fossil fuel energy rather than the environmental impacts that we recognize now, his main message was about the need to curtail the rapid increase in energy use. Historically speaking, this was an extraordinary idea. Carter put words to the concept that, from a resource perspective, the world was turning a corner.

Carter spoke of sacrifice for the sake of future generations, a logical strategy considering that he labeled the challenge "the moral equivalent of war," and he was himself of the World War II generation that pulled together for victory over fascism. By the late seventies, however, the Boomers were turning thirty and were not interested in sacrifice. Carter's message galvanized a few to action who would begin shaping new energy policy concepts, researching building methods and materials, and inventing energy-efficient technologies. For the most part, however, Carter's words fell on indifferent ears. When Carter left office, federal support for his energy campaign collapsed, Reagan removed the solar panels from the White House, and the urgency to become a less energy-intensive society receded to the back of the public mind.

This silently emerging energy crisis was taking place against a complex social and political backdrop. The nation had just endured the militarily unsatisfactory and domestically divisive Vietnam War and, in 1973, found itself in the midst of the Watergate crisis. Internationally, the Cold War popped up in the Middle East as Western nations aligned with Israel and the Soviet Union with Israel's Arab neighbors in the continuing conflict over Israel's right to exist as a nation. It was the Arab-Israeli confrontation that brought about the 1973 embargo by nations belonging to the Organization of Arab Petroleum Exporting Countries, or OAPEC. Because we were importing less than 10 percent of our oil from Arab nations, the embargo might have had less psychological impact on the American public had a strong policy been put in place to deal with the gradually fraying U.S. energy infrastructure. As it was, the 1973 oil embargo served as a catalyst for bringing U.S. energy vulnerability to everyone's attention.

Founded in 1968, OAPEC's original purpose was to limit the influence of radical regional politics on the international petroleum business. The conservative OAPEC membership included only Kuwait, Libya, and Saudi Arabia. However, by 1972, membership qualifications had changed and the more activist nations Algeria, Iraq, Syria, and Egypt were admitted. The temptation to play the oil card became too much with the October 1973 outbreak of the Yom Kippur War, the aggression of Syria and Egypt against Israel to regain land Israel seized during the 1967 War. OAPEC members, along with Arab members of OPEC (Organization of Petroleum Exporting Countries), which included Iran, wanted to punish Israel's primary allies and began cutting exports of oil, primarily to the United States and the Netherlands. They continued to do so for five months until an agreement with the United States could be reached.

While the roots of the embargo lay in the highly complex events surrounding the legitimacy of the Israeli State, the result for the Arab oil nations was that they realized the political power they held through their oil reserves and recognized their opportunity to hike the price of oil to reflect its true market worth. Shock waves from the embargo and the resulting upward shift in world oil prices included recession, inflation, and higher unemployment in the United States. The 1973 oil crisis proved to be a wake-up call about energy sources in general for state and local governments, utilities, and forward-thinking industries. Beginning in the late seventies and early eighties several things began to happen in response to the oil embargo and other emerging energy issues that would have a profound effect on energy use in the United States.

It was the end of low prices for Mideast oil and the beginning of energy insecurity in the United States, as control of energy commodities slipped into a wider, international arena. The federal government lowered the speed limit nationally to 55 miles per hour, incidentally saving lives as well as gasoline. In 1975, President Ford signed the Energy Policy and Conservation Act (EPCA), establishing the Strategic Petroleum Reserve (SPR), which currently holds over 775 million barrels of oil in underground caverns along the Gulf of Mexico. The SPR serves primarily to protect U.S. interests against future use of oil as an international political football. And Title V of the Energy Policy and Conservation Act of 1975 established the Corporate Average Fuel Economy standards (CAFÉ) setting lower limits on miles-per-gallon vehicle efficiency. Americans started buying smaller cars, most of them foreign imports with better gas mileage than American models.

Federal Lawsuits and a Network of State Energy Offices

Meanwhile, in a parallel political universe, the federal judiciary was engaging in a series of lawsuits against domestic oil producers who had taken advantage of the oil crisis to overcharge consumers. These companies were ordered to make restitution through the Petroleum Violation Escrow (PVE) funds, which essentially built a national network of state energy offices. Also known as "Oil Overcharge Funds," this money promoted a variety of energy conservation and efficiency programs and indirectly assisted national efforts through local partnerships at the state level.

When the Arab oil-producing nations cut off their exports to the United States, the U.S. Government put oil pricing regulations into effect from 1973 to 1981 to prevent price gouging on the part of domestic oil producers. The U.S. Department of Energy (U.S. DOE), founded in 1977 to coordinate and centralize the various energy-related federal functions, was charged with keeping an eye on violators. In 1981, the first lawsuit was settled against Chevron. Judgment in this case took the form of indirect settlement, which became the pattern for future cases.

Indirect settlement meant that the fines paid by the oil companies for this and subsequent court actions would be divided up among the states based on volume of petroleum use in each state, and administered by state governments to carry out specific federal programs. In addition to the oil company lawsuits, PVE funds were also made available through legislation in The Warner Amendment to the Further Continuing Appropriations Act in 1983. The first settlement with Chevron involved a total of $25 million, and up through the last settlement in 1986 with Diamond Shamrock, over four billion dollars were distributed to the states.

The indirect settlement concept was a stroke of genius because it not only solved the problem of repaying a massive group of consumers for being overcharged on gasoline, it created the network of state energy offices that was key to much of the progress that has been made over the last thirty years. When the funds were first distributed to the states, each state was mandated to set up an energy office to administer its portion. Some states chose to locate the energy office within their natural resource departments, others in their economic development offices. The federal settlements outlined specific activities and programs for which the states could use the money.

The Chevron Settlement specified that states could use funds for promoting ride-sharing and public transportation, performing building energy audits, providing grants or loans for weatherization and equipment installation, along with funding a variety of transportation infrastructure improvements. When the Warner Amendment was signed two years later, it created the formal programs that would define the Oil Overcharge funding up to the present day. These included the State Energy Conservation Program (SECP) that offered a variety of grants for efficiency and renewable energy projects, the Institutional Conservation Program (ICP) that provided matching grants for energy efficiency improvements to schools and other public buildings, and the Energy Extension Service (EES) that made use of existing state agricultural extension networks to distribute information to the public about energy efficiency and renewable energy.

These programs are no longer running, but the federal government still provides funding for similar activities through the State Energy Program (SEP), which grants funding to states for specific energy projects. U.S. DOE's Weatherization Assistance Program (WAP) and the Low-Income Energy Assistance Program (LIHEAP), administered by the U.S. Department of Health and Human Services, also receive funding from the federal government, but funds are distributed by the states. State energy offices now follow their own drumbeat with a little help from U.S. DOE, and the national network is much more loosely defined than it was.

Even though most of the Oil Overcharge Programs are no longer in place, the lawsuits of the eighties gave us a practical structure for working on energy priorities at the state level that still has use today. In the seventies, when we were reacting to the oil embargo, our primary concerns revolved around reducing dependence on foreign oil and seeking out other domestic sources of energy to increase energy security. Environmental concerns did not enter into it, and were even considered totally unrelated to the energy issues we were confronting. Today, states are working on carbon emissions reduction plans and quantifying their bioenergy resources for economic development, and state energy offices find themselves in a new role, serving environmental priorities but still promoting many of the same energy concepts and technologies.

Appliance Efficiency Standards

Before the oil embargo, use of the phrase "energy efficiency" was confined to the world of utility and process engineering that attempted to squeeze more work out of equipment while using the same amount or less fuel. For example, utilities were able to reduce the cost of electricity substantially over the years by making generation equipment more energy efficient. Thomas Edison's first power station on Pearl Street in New York City sold electricity for about 25¢ for one kilowatt hour. After 125 years of inflation, that would compare to around $4.50 per kilowatt hour in today's currency. Most of us now pay between 8¢ and 15¢ for one kilowatt hour.

Gains in generation efficiency leveled out in the 1950s, but engineers have continued to improve the efficiency of other equipment that uses electricity, from home appliances to heavy industrial process equipment. The opportunity to make consumer products more efficient was not lost on Washington. The most successful federal strategy for reducing energy use has been mandating efficiency standards for appliances and building mechanical equipment. Appliance and equipment efficiency standards have had bipartisan appeal. Beginning with President Reagan in 1987, four presidents have signed legislation mandating increasing levels of energy efficiency for household and commercial appliances and for heating and cooling equipment. These standards have been very effective.

The federal government adopted energy efficiency standards for appliances for a number of reasons. Naturally, reducing energy use and peak demand were important considerations, but the effort was also meant to deal with certain market failures that had created disadvantages for consumers. Renters and new home buyers were at the mercy of landlords or builders who purchased and installed appliances without regard to the cost of operating them. Information about appliance performance was often misleading to potential purchasers. Manufacturers of efficient appliances priced their products like luxury items because of their low volume of production. A number of states had adopted standards for appliance efficiency but they were inconsistent from state to state, leaving manufacturers caught in the middle.

By 1986, energy efficiency advocates and appliance manufacturers agreed it was time for a consistent national standard for certain products and they threw their effort and support behind the National Appliance Energy Conservation Act (NAECA), signed by President Reagan in 1987. This legislation established a process for setting standards that weigh the costs and benefits for both manufacturers and consumers and was designed to be updated from time to time as greater efficiencies became practical and economical. Product standards in this legislation were for residential room air conditioners, water heaters, refrigerators, freezers, dishwashers, clothes washers, dryers, ovens, and ranges. The 1992 Energy Policy signed by President George H. W. Bush added some commercial and industrial equipment such as electric motors and mechanical equipment, along with lighting equipment. President Bush also upgraded residential appliance standards in 1989 and 1992, and the Clinton Administration raised them again, adding central air conditioning and heat pump standards. The second Bush Administration has confirmed standards for clothes washers and water heaters set by the Clinton Administration.

It's hard to picture energy we haven't used, but the savings that have resulted from the appliance standards have helped us avoid building many new power plants and transmission lines. Also, these efficiency standards have made using less energy quite painless. And savings from efficient product standards will continue to accrue as old equipment is replaced over time. According to the American Council for an Energy Efficient Economy (ACEEE),[1] cumulative savings from existing standards will have saved 6.5 percent of projected electricity use, or 250 billion kilowatt hours, by 2010. They project that from 1990 to 2030, U.S. residential and business consumers will save approximately $186 billion in energy costs as a result of using more efficient equipment.

ACEEE also recommends that future updates in the efficiency standards include products that have not appeared on the list before, such as torchiere lighting fixtures, traffic lights, illuminated exit signs, commercial refrigeration equipment, vending machines, and electronics that still use electricity when they are turned off, creating what is called a "phantom load." ACEEE projects that, by raising the efficiency standards of these products, plus building transformers, commercial unit heaters, residential furnace fans, and ceiling fans, the energy savings in 2020 will amount to 5 percent of projected energy use for that year and would reduce peak electrical demand by about two hundred 300-megawatt power plants. The federal efficiency standards have been very successful already, but they remain a useful tool for achieving far more savings in the future.

ENERGY STAR®

Another resounding federal success is the ENERGY STAR program. ENERGY STAR was started as a voluntary program in 1992 by the U.S. Environmental Protection Agency and was the first federal effort to promote energy-efficient products for environmental reasons. The ENERGY STAR qualification is like the whipped cream topping on the cake because its label appears only on products that meet the program's own high standards, which are higher than the legislated minimum efficiencies. For example, higher efficiency standards for refrigerators mean that any unit bought today will be at least 20 percent more efficient than one made before the current standards were adopted. An ENERGY STAR–qualified refrigerator must be at least 40 percent more efficient. The program also maintains standards for products that have not yet been included on the national standards lists. The first products to earn the ENERGY STAR were computers and monitors, and the familiar label is now on a broad variety of products from washing machines to new homes. U.S. EPA partnered with U.S. DOE in 1996 to broaden the scope of ENERGY STAR–qualified products.

The ENERGY STAR program is still voluntary and continues to attract a following from both businesses and consumers. ENERGY STAR is now a highly recognized and trusted brand. It can claim partnerships with over 9,000 manufacturers, retailers, builders, commercial businesses, utilities, and government bodies to promote, not only efficient products, but building construction methods and energy management procedures as well. There are over fifty categories of ENERGY STAR–qualified products and more are sought each year. The program is proactive in finding specific types of products such as compact fluorescents for torchiere lamps and energy-efficient vending machines. Their building efficiency programs include partnerships with new home builders and with state and local organizations for the Home Performance with ENERGY STAR program, which focuses on improving the efficiency of existing homes. More recently, they have set standards for commercial construction and industrial processes. The ENERGY STAR program is also an ambassador of good will as well as efficiency. It has established a working relationship with European Union nations to coordinate standards for efficient office equipment, which has an international market.

Utility Action

Concern about the 1973 oil embargo prompted a different response from state government regulatory agencies and the utilities they regulated. The tightened oil supply began to drive up the price of electricity for utilities using oil or gas. The public, which had enjoyed many years of cheap electricity as the utilities had built increasing economies of scale into their networks, began to complain. Regulations to control power plant emissions were in the works, and people became concerned about potential waste and safety hazards of nuclear plants, which the Three Mile Island accident in 1978 only confirmed. The oil embargo was the tipping point for a heightened awareness about energy resources in general, which prompted a reassessment of the basic structure of the electric utility industry. Over the next few years, dramatic changes began to occur.

For several decades, electric utilities had enjoyed essentially monopoly status, each with its own service territory franchise, granted by the state regulatory agency. The regulators then negotiated rates to assure utilities a fair profit while keeping rates low for consumers. This arrangement worked well for both the utilities and their customers for many years, because the utilities could operate as private companies, but the service they provided was regulated for the public good. As long as they could continue to grow while keeping rates low, this monopolistic structure seemed to be the most cost-efficient way to deliver electricity. But by the seventies, and with the boost in oil prices from the embargo, this system began to crack around the edges as the industry ran up against the limits of its own economies of scale.

Substantive changes began at the national level. Recognizing that perhaps the electric utility industry could use a little competition, the federal government passed the Public Utilities Regulatory Policies Act (PURPA) in 1978, requiring utilities to purchase power from nonutility generators when it was offered at the same cost utilities would pay to generate it themselves. PURPA would open up opportunities for new players and would begin to nudge the industry in the direction of deregulation. Utilities were not pleased with the idea of giving up their long-time control over generation.

Another federal law passed in the same year, the National Energy Conservation Policy Act (NECPA), included the requirement that utilities offer home energy audits to their customers to promote energy efficiency. This essentially introduced a second new source of economical electric power—the saved kilowatt hour. To say the least, these two laws made life more complicated for utilities, although some saw the opportunities present in encouraging their customers to be more efficient and had already begun to investigate the possibilities. Things also became far less clear-cut for the state regulatory agencies, and tensions increased between utilities and regulators. The collaborative relationship the two parties had shared for many years under the monopolistic franchise system became contentious and distrustful and one that was more clearly the interplay between the regulator and the regulated.

Demand side management (DSM) emerged from the changing dynamics in the utility industry and was perhaps the most positive product of these times. State regulators picked up the cue from the National Energy Conservation Policy Act and expanded requirements that utilities develop programs to help their customers reduce both energy use and peak load. DSM was part of Integrated Resource Planning (IRP), a requirement by regulators that utilities consider the economics of all fuel and energy resources when planning new generation, rather than continuing to rely on the convenience of past sources regardless of cost. The original purpose of demand side management was to reduce or delay the need to build new power plants. Once in place, DSM programs played well with utility customers who were beginning to notice the visual and environmental intrusion of power plants in the landscape. Over time, DSM programs took on a marketing function as well because they conveyed an image of utility community service to both regulators and customers.

What Is DSM?

DSM, or demand side management, simply means managing the use of—or demand for—energy, as opposed to supply side management, which looks at obtaining an adequate and economical supply to sell to customers. For utilities, demand side management involves working with their customers to influence how much energy they use or when they use it. Demand side management in the old days would have meant simply encouraging people to use more electricity to support the growth of the utility infrastructure, and that was essentially what utilities did. However, as regulators began to require demand side management as part of an integrated resource plan, utilities developed their own programs to help customers in all sectors, relying on several strategies for both reducing energy use and shifting use to off-peak hours.

For residential customers there were bill stuffers and other publications offering ways to conserve energy, along with other information activities like booths at public events or collaborations with the state energy office on energy efficiency workshops. Residential customers were also offered home energy audits and financial incentives or low-cost loans to replace old, inefficient appliances like refrigerators or water heaters. Other strategies were even more effective with business and industrial customers, whose rates are structured differently because of higher usage. Performance contracting is one example. An energy efficiency contractor agrees to install energy-efficient equipment at a utility customer's facility at their own expense, and is then paid back with the resulting savings. Performance contracting has become a major energy efficiency industry and is warmly embraced by municipalities and public institutions like universities who are always scuffling for capital improvement dollars for their buildings, but who want to save on energy expenses.

For industrial customers, load shifting methods are often effective. Utilities might offer a cash incentive to a manufacturer to move a certain energy-intensive process out of peak hours. There are also special rates offered depending on the time of day, which can accomplish the same goal. Then there are programs that charge a lower electricity rate to customers willing to be shut off if the utilities peak load gets dangerously high. These might be industrial customers with production flexibility during the summer when local cooling loads during peak hours are high.

Many utilities have found DSM programs useful to their bottom lines. Large commercial or industrial customers can still have significant impact on utility load management by reducing or moving their load around. Demand side programs remain popular with utility customers from all sectors, making them an important public relations tool. DSM programs nationally have created an energy services industry of energy auditors, weatherizers, equipment and controls manufacturers and installers, and other energy professionals whose jobs serve the utility programs. These programs have also had a side benefit of connecting the investor-owned utilities more closely with their communities through collaboration with local groups and sponsorship of a variety of events such as fairs and environmental conferences. Not all private utilities take advantage of this opportunity, but those who do, usually the smaller, local utilities, often enjoy the kind of relationship with their community usually associated with a municipal or cooperative utility.

The decade of the eighties brought DSM programs to their peak as most state regulatory agencies adopted some form for Integrated Resource Management requirements. By the late 1990s state regulatory agencies had begun to eliminate these requirements, and the implementation of DSM programs fell dramatically. The role of demand side management programs in the industry picture has remained an important one but has changed somewhat since the early regulatory days. As the industry heads toward deregulation, the question naturally arises as to whether market forces, which cannot always be relied upon to act in the interests of the common good, will promote the advances in utility energy efficiency we need to slow global warming.

Public Benefits Programs

One approach to keeping DSM efficiency programs strong is to collect funds from utility customers to pay for them. As this places some of the responsibility for the environmental impacts of generation on energy users, utilities find this a more equitable arrangement than stringent regulations that hold them fully accountable. A number of states (primarily in the Northeast and upper Midwest) have already established "public benefits" programs where utilities are required to collect a small amount from each customer, based on usage. This money is used to promote energy efficiency and renewable energy through programs similar to DSM. In some states part of the funding is spent for research exploring the environmental impacts of energy use.

Public benefits programs have adopted some of the more popular DSM efforts, such as publishing energy information and giving financial incentives for buying more efficient water heaters or refrigerators, or installing solar electric panels. The structure of these programs varies from state to state, with some being administered by state agencies and others set up to keep the funding separate from the state's general fund. This type of program is certain to attract more states as time goes by because it can be an effective way to deliver energy services economically and support the growing energy efficiency industry through time-tested DSM methods.

Building Energy Codes

Demand side management programs address different sources of energy savings than appliance and equipment standards, and together they cover a lot of ground. However, the area of building and home construction presents a whole different array of opportunities to save energy. Even though the ENERGY STAR program has developed standards for residential and commercial construction, these standards remain voluntary. Mandatory building energy codes for both commercial and residential construction have been widely adopted by both state and local governments, although enforcement is uneven.

When we hear about something being "built to code," the code in question is the set of codes adopted by state or local jurisdictions to mandate minimum requirements for building construction and electrical, mechanical (heating, cooling, and ventilation), and plumbing systems. These codes are designed to protect public health and safety, and they specify such things as design of systems and construction details, quality of materials or equipment, and installation and construction methods. These codes have not traditionally included anything about energy efficiency or use. Code standards to promote building energy efficiency have been developed independently, and once again, the oil embargo of 1973 was the inspiration.

Like other building codes, energy codes are written to be used by architects, engineers, equipment manufacturers, builders, and code officials. They must therefore be clear about what is required, and they need to be enforceable. Energy codes set minimum requirements for the building envelope (walls, windows, roof, and floor), the mechanical system, and in commercial codes, the lighting. Codes only establish minimum levels for compliance, which means that optimum energy efficiency is usually not achieved by following the code exactly. The codes don't necessarily include all the measures that would contribute to the efficiency of a specific project. For example, site related design strategies like building orientation that uses solar energy or creates natural ventilation, or planting trees to increase shading, are not included. However, cost effectiveness of measures is considered to make energy code requirements more acceptable to builders and buyers. Cost effectiveness is measured by how quickly the energy savings from a particular measure will pay for any additional investment.

There are two energy code standards that have been generally adopted by state and local governments across the country. The ASHRAE 90.1 standard is normally adopted, or serves as the model for many local commercial building energy codes. The International Energy Conservation Code (IECC) is most often chosen for residential buildings, although their commercial building model code is frequently adopted as well, and actually references the ASHRAE 90.1 standard.

ASHRAE 90.1

Shortly after the oil embargo, the American Society of Heating, Refrigeration and Air Conditioning Engineers (ASHRAE) began developing a comprehensive energy efficiency standard for commercial and other non-residential buildings, as well as high-rise residential construction. ASHRAE members are primarily mechanical engineers whose job it is to design commercial-scale mechanical and refrigeration systems. The organization was fully aware of the opportunities for reducing energy use in large buildings and immediately set to work. In 1975 they issued ASHRAE 90.0, which was then extensively revised and reissued as

ASHRAE 90.1 in 1989. The standard has since been updated several times and is referred to by year. ASHRAE 90.1 has become the industry standard for commercial and other nonresidential buildings. ASHRAE developed their 90.2 standard for residential buildings in 1994. This standard has not been widely adopted because most jurisdictions had already adopted the IECC.

The International Energy Conservation Code (IECC)

The International Energy Conservation Code is one of several model codes developed by the International Code Council (ICC). Other International Codes, or I-Codes, include the International Building Code, the International Fire Code, and the International Zoning Code. ICC membership is formed of building and fire code officials, architects, engineers, builders, elected officials, and manufacturers, among others. Three regional code organizations joined forces to form the national ICC in 1994 when it seemed like a good idea to come up with a set of national standards. Previously, each of the three organizations had their own model code based on regional climate requirements. The ICC uses a consensus process to develop its model codes, which allows input from all members. The Council's voting members, who are all code officials, make the final decisions in the public interest. The IECC was originally called the Model Energy Code and was created by the Council of American Building Officials, one of the three regional organizations that formed the ICC. The IECC includes minimum energy efficiency provisions for both commercial and residential buildings. The commercial building standard incorporates ASHRAE 90.1. Like ASHRAE 90.1, the IECC is periodically updated.

Both ASHRAE 90.1 and the IECC are model codes, which means that they are not laws themselves but can be adopted by a city, county, or state jurisdiction as a legal code. Jurisdictions can change or add provisions as they like, but the heavy lifting has already been done by the national experts. In 1992 the federal government passed the Energy Conservation and Production Act, which requires states to certify that they have adopted a building energy code that meets or exceeds ASHRAE 90.1. Most but not all states have done this. Considering that buildings consume between 35 and 40 percent of the nation's energy, we need an energy code structure that is consistent, effective, and enforced. The model codes provide a strong foundation, but plenty of work remains, particularly in educating code officials to improve enforcement.

The Green Building Movement

Environmental concerns go beyond energy, but there is considerable overlap of the issues when designing and constructing buildings or urban spaces. Now that the environmental impacts of energy use are more fully understood, we are moving beyond simply improving the efficiency of our structures and equipment. "Green" or sustainable building practices, as currently defined by several progressive programs, are moving us in that direction by integrating energy considerations into the overall design process.

The concept of green building actually began in the seventies, although it would take almost thirty years for the idea to be embraced by the general public. When the energy crisis hit the headlines in 1973, we were mostly worried about where our energy supply was coming from, not about any connection that energy use might have to environmental quality. It is therefore not surprising that federal appliance standards and building energy codes were the first strategies to emerge, as they were both aimed totally at reducing energy use. Even though the ENERGY STAR program originated in the U.S. Environmental Protection Agency, the government promoted it as an energy reduction program, linking it to improving air quality by reducing power plant emissions associated with acid rain.

Even so, the environmental movement was beginning to make some big connections between profligate resource consumption and reduced environmental quality. A small niche market began to grow for homes that used less energy and were efficient with other resources as well. Many of these homes were owner-built, and those owner/builders with the necessary skills and passion became specialty builders, commissioned to construct elaborate, handmade homes for others from recycled materials, straw bales, adobe, or stone. They frequently used solar energy for electricity and hot water.

Alternative Construction Was the First Green Building

Experimental builders were exploring traditional building techniques like straw bale construction, rammed earth, cob (a mixture of straw and mud that can be used to build freestanding walls), cordwood construction, and adobe bricks. Architect Steve Reynolds of Taos, New Mexico, invented the Earthship, a self-sufficient home design built of old tires filled with sand that included a greenhouse for growing food, water harvesting from the roof, composting toilets, and solar energy for electricity and hot water. Alternative construction techniques may have varied, but there were certain qualities they held in common. First, the materials came from local sources, or as close to the site as possible. Also, the materials were to be natural unless there were discarded items that could be gracefully recycled, like plumbing fixtures, windows, or the old tires in the Earthships, which were plastered over and not visible when the home was finished. Many builders were also concerned that the materials used for construction be nontoxic to the inhabitants.

Alternative builders considered energy use carefully in their designs, drawing on the wisdom of traditional building techniques indigenous to their area. They oriented the homes to make use of the sun and of ventilating breezes. Some were buried in hillsides to make them easier to heat. Some were underground altogether. They incorporated logs and tree parts as structural decoration, or they were carefully and precisely crafted with old fashioned post and beam construction. Many of these homes were built primarily with sweat equity in rural areas away from power lines, gas lines, and municipal water. Budgets didn't permit a huge solar electric array, so appliances and lighting were minimal. Energy and water efficiency were essential. If an alternative builder was hired by a client with money to spend, these clients tended to be people for whom this simple and natural approach to energy use was also appealing.

A subculture began to grow among people who were reconsidering the way homes are constructed. They felt that most people had lost touch with how their own homes were built and had forgotten what American traditions of home building and barn raising were all about. This alternative construction movement possessed a strong sense of personal and community involvement in producing shelter for people. The homes themselves, including most interior woodwork details, floor, and wall finishes, were hand crafted with care. Often, much of the work was accomplished by family members of all ages along with friends and strangers drawn in to learn how to build their own. The design and construction of one of these handmade homes integrated natural and local materials, low embodied energy as well as efficient energy use, and indoor environmental comfort. The builders were committed to a strong environmental ethic and to the idea that everyone, regardless of economic status, should be able to build a healthy and resource-efficient home. Many of the early practitioners are still in business, continuing to offer both construction services and instruction to a new generation of like-minded home builders.

Although it was unable to catch the attention of the typical mainstream home buyer, the alternative construction movement contributed a great deal to what we now call "green" building. During the eighties and nineties, as these custom green homes became more sophisticated, the market began to attract manufacturers who would provide natural paints, finishes, and other materials. Flooring made of sustainably harvested wood, cork, and bamboo became available. Alternative builders gave workshops and traded ideas and techniques. Gradually a network of alternative and natural construction resources grew and a few mainstream architects and builders began to notice. In 1991, Austin, Texas, launched its Green Building Program, the first in the country. The Austin City Council and its municipal electric utility, Austin Electric, had incorporated ENERGY STAR into its program for new homes in 1985 to postpone the need for a new power plant. It soon transformed into the Austin Green Building Program, which included additional resource sensitive techniques such as water conservation and passive solar orientation. In 1995 Austin Electric introduced a commercial building component as well.

Today the program offers technical assistance and workshops and publishes a comprehensive *Sustainable Building Sourcebook*. The sourcebook reflects the values of that early alternative construction movement, containing such criteria as maintaining the natural integrity of the site, recycling and reusing building materials, employing active and passive energy efficiency strategies, creating a healthy indoor environment by using nontoxic materials and finishes, and conserving and reusing water. These categories are all essential to what building professionals today call green or sustainable building techniques.

Built Green Colorado is another successful green building program that was founded in 1995. It has since become the largest residential green building program in the country. The city of Denver, Colorado, joined with the local homebuilders association, the governor's office, and the area utility to establish a green homebuilding program to serve its metropolitan area. This is a voluntary program for builders who agree that they will build homes that are energy and water efficient, that they will use building materials efficiently, and that they will improve indoor air quality for the occupants. Their homes must also be durable and economical to maintain.

The alternative builder's first priority was the home's environmental impact, followed closely by the physical and spiritual health and comfort of its inhabitants. Time and money enter in only as necessary to accomplish the first two. Idealistic as this philosophy might be, the alternative home is really what many of us would like to live in. We would love to see the essence of the alternative approach translated to a replicable mainstream green building model that would make such homes available and affordable for people nationwide. Both Built Green Colorado and the Austin Green Building Program continue to inspire other cities searching for ways to encourage green residential construction.

Even though only a small percentage of American homes could currently be called green as defined by these or other programs, residential builders are still ahead of the commercial sector. Very few commercial and other nonresidential buildings can claim to be green. Commercial buildings represent the opposite end of the construction industry spectrum from the alternatively constructed, handmade home. Commercial projects are generally big and frequently costly, built on tight schedules for occupants who are often not the building owners. Many buildings suffer from "sick building syndrome," with poor indoor air quality that ends up costing employers a great deal of money in sick days and turnover. There is also much energy wasted in most commercial buildings in lighting and mechanical systems, as well as from poor insulation and windows. Until the last decade or two, there has been little evidence to suggest that indoor air quality and energy use should be of concern for commercial building owners who were more worried about construction costs per square foot. Architects and commercial builders have adhered to the owner's priorities of budget and schedule, knowing that trying something new takes time and costs money, neither of which attracts new clients.

The LEED® Standard

Fully conscious of the bottom line priorities of the commercial building industry, the U.S. Green Building Council (USGBC) was formed in 1993 to promote sustainability in the commercial construction industry by focusing on environmental responsibility, cost effectiveness, and occupant health. The organization comprises industry professionals, including architects, builders, real estate developers, facility managers, manufacturers, government agency staff, and nonprofit organizations, who belong to seventy regional U.S. chapters. Its vision is to raise the bar for mainstream commercial building practices to a level of high-performance green building. Its rating system, the Leadership in Energy and Environmental Design (LEED®), has caused quite a stir within the industry, but more important to its long-term success, it has caught the public imagination. Cities and towns, universities, government agencies, nonprofit organizations, and private developers are all intrigued by the idea of building a LEED certified building or of mandating a LEED standard for all future buildings.

The USGBC is moving toward its goal of transforming the building industry by stimulating consumer demand with the promise of a LEED Gold, Silver, or Platinum–rated building. USGBC has developed standards for new construction (LEED-NC), existing buildings (LEED-EB), commercial interiors (LEED-CI), core and shell (LEED-CS, buildings like shopping centers with multiple tenants who finish their own spaces) and is soon issuing a standard for homes and for neighborhood development. According to the Council's June 2007 fact sheet, there are close to 900 LEED-certified buildings in the United States and 6,800 registered LEED projects.[2] These are small numbers compared to the 170,000 commercial buildings built each year. However, the fact sheet also points out that there are LEED projects in all fifty states, and twenty-two states and seventy-four local governments have adopted the LEED standard as part of their sustainability or carbon reduction plan.

The LEED process is a team-oriented, integrated approach to design that addresses five basic areas of concern. Once again, these are reminiscent of the alternative home builder's list: sustainable sites, water efficiency, energy and atmosphere, materials and resources, and indoor air quality. Each of these categories includes a number of very specific criteria with points assigned to them. A project could earn a point for building on a previously developed site or for reusing part or all of an existing structure on the site. There are points for using recycled or local materials or for including showers and bike racks for bicycle commuters. The total number of points earned determines the level: Certified, Silver,

Gold, or Platinum. Recordkeeping for a LEED project is a labor-intensive process because the whole point of earning the coveted certification is to guarantee that the building is performing as promised. The project team must keep track of how it meets the requirements for each point. For example, in order to earn the point for using nontoxic paints and finishes, they must show statements from the paint manufacturer that its product passed the recognized standard adopted by LEED.

LEED certification continues to evolve. Recent concerns about the reduction of carbon emissions and the need to reduce energy use in commercial buildings are reflected in revisions to the way a project is scored. Originally, it was possible to earn LEED certification by focusing on nonenergy-related points. Now, required levels of energy efficiency have been raised for all projects.

The buildings that have been certified are beautiful demonstrations of green building principles. Many of them are public buildings like libraries and nature centers. Many are owner-occupied corporate or organizational headquarters, serving as reflections of the owner's commitment to environmental quality. The USGBC wants to transform the industry by making high-performance green building the new construction "business as usual." When a municipality or a state government decides that all new government buildings will be built to LEED standard, it is a sign that the transformation has begun. Another indication is the 2007 Memorandum of Understanding between the USGBC and the International Code Council (ICC) to work together in promoting green building principles.

Costs and Benefits of Green Building

Green or sustainable building practices are large steps forward toward using energy effectively. Energy use is not only addressed in Energy and Atmosphere, it is woven throughout the LEED categories, from providing transportation options for building employees to the embedded energy in recycled materials. LEED, ENERGY STAR, and other green building programs are slowly turning the giant building industry to move in a new direction. However, the main argument from those who continue to resist is cost. Many builder and developers claim that adding green features to a building makes it too expensive. It's true that initial costs are greater in most cases, but it's also true that energy savings alone usually pay back the difference within a few years if not a few months. Another point to consider is that the green building industry is new and just learning about what works most effectively in meeting these new priorities. Extra costs will go down as we achieve economies of scale and experience.

The idea that the cost of green buildings is significantly higher than standard construction is something of a myth. The World Business Council for Sustainable Development published results of a 2007 survey of building industry professionals in eight countries, including the United States.[3] The survey included the full spectrum from architects, journalists, and regulators to developers, engineers, landlords, and corporate tenants. There were two questions related to the participants' knowledge of sustainable building issues. When asked what the cost premium for a green building would be as a percentage of the total project, answers from Western nations ranged from 12 percent (French respondents) to 19 percent from Spanish respondents. Americans put the green premium at 16 percent, and the international average estimate was 17 percent. The report states the actual average green premium to be under 5 percent in Western nations.

The second question asked for an estimate of the greenhouse gas emissions that come directly or indirectly from buildings. The international average estimate in this case was 19 percent, with the lowest being the U.S. guess at 12 percent and the highest from Spain at 30 percent. In actuality, buildings contribute about 40 percent of greenhouse gas emissions to the atmosphere. It is interesting to note that this was not a survey of general public knowledge, but rather a survey of professionals most closely associated with buildings and the building industry.

Net Zero Energy

Green buildings will get us closer, but arriving at our goal of truly sustainable energy will mean moving all the way to net zero-energy homes, businesses, and communities. This means that not only will energy efficiency be integrated into sites and structures, the remaining demand for energy will be met at the site or nearby. We already have the technology to drastically reduce energy use in homes and buildings. The remaining energy requirements could be met with electricity and heat from renewable sources. This does not mean that each structure would be producing all its own energy all the time. There would still be an electric grid that homes and other buildings would use like a rechargeable battery, sometimes putting electricity in and sometimes taking it out. Over the course of a year, the result would be "net zero." Local energy production could extend to neighborhood or campus generation plants, or district heating and cooling systems that use waste heat or biomass sources. Money spent on energy would stay in the community, local energy businesses would create local jobs, and opportunities would open up for small investments in the local energy infrastructure.

The U.S. Department of Energy runs a program called Building America that does research on construction methods for high-efficiency production homes and on improving existing housing stock. Their ultimate objective is to find the most cost effective way to build zero-energy homes on a large scale. Building America is structured as a public/ private partnership, bringing together five industry partnership teams across the country and involving almost 500 companies. The idea is to encourage the cross-pollination of ideas among architects, engineers, equipment manufacturers, builders, planners, material suppliers, financial institutions, and contractors. The program uses a systems design approach to home building, integrating all possibilities for saving energy, from the envelope to the mechanical system. The resulting homes are rigorously tested for their performance. According to the U.S. DOE, homes built through this program are performing at 40 to 70 percent greater efficiency than standard production homes.

Meanwhile, a variety of developers, nonprofits, and utilities in different states are going ahead to build experimental zero-energy homes with varying degrees of success. Obviously, initial costs are higher than standard construction at this time, particularly because of the price tag for an electric generation system. However, if the home is truly "net zero energy," or even close to that, there will be next to no utility bills and considerably lower carbon emissions. The extra cost will pay back fairly quickly through the energy savings, but the environmental advantages begin immediately. The concept of net zero energy as the way to build represents an entirely new way of looking at energy delivery and use. We need to get used to the idea that we can take on the responsibility for producing our own energy through dramatically reducing demand and generating the rest of what we need ourselves, either individually or within our communities.

Energy Conservation: The Essential Challenge

From efficient appliance standards to energy codes and green building, the possibilities for using energy more efficiently have been fully and scientifically explored. Increasing energy efficiency is a design problem, whether we are talking about a more efficient furnace or a public transit–oriented neighborhood. Whatever specifications, codes, or regulations are required to get the job done can be instituted, and the results verified. However, the amount of energy we choose to use in our homes, businesses, or vehicles remains a wild card. Energy consumption is so intimately tied to the American consumer culture that it is impossible to make energy a separate issue.

It will not be easy for Americans to come to terms with increasingly limited resources, because America has never been about limits. The original American Dream was about profiting from your hard work and having control over your own life, property, and money. The right of every person to prosper, regardless of social or national origin, was woven into the complex fabric of American idealism from the beginning. Coal and oil provided the fuel for America's awesome growth. Indeed, the United States owes its exceptional material wealth to the happy coincidence that within its borders reside immense forests, rich farm land, and the conduit of mighty rivers, in combination with significant amounts of coal and oil to develop them. And who better to make this happen than the many groups of energetic immigrants fleeing political and religious persecution, repression, and poverty, ready to roll up their sleeves and harness their ingenuity to build a better life?

It is not surprising that today, within the context of our singular sociopolitical history and our nation's natural wealth, Americans have come to feel a certain entitlement to abundance and the unlimited right of acquisition. This contemporary American value is now colliding with the limits of the earth's resources. The long debated question of individual rights versus the common good has direct bearing on the essential resource questions we are facing. Decoupling our consumption habits from our national identity is an important first step toward significantly reducing the ecological footprint of American society. Rampant consumerism is not a democratic value, nor is it one we should be proud of exporting along with the ideals of democracy that many other nations would emulate.

Energy and Our Stuff

Consumerism could be defined as a way of life where you are forced to buy more and more stuff in order to keep up with what you've already got. We have built an economy that depends on the marketing adage to "create a need and fill it." Anyone who can remember record albums and stereo system components that were then replaced by 8-track tapes, then cassette tapes, and now CDs, and next, internet downloading, will remember not only replacing equipment but buying whole collections on a new recording medium in order to continue enjoying the same music. Existing products become obsolete as new products emerge, requiring periodic retooling of the household. Upgrading a computer is a case in point. Getting something fixed is sometimes impossible and usually more expensive than buying a new one. What isn't necessarily obvious is that energy is the foundation of our consumer culture because it is the means for both producing and using all these products. Fossil fuel energy has been relatively cheap. Inexpensive plastics made from petroleum mean that buying more than one television set per household is easily affordable. Cheap electricity means the monthly bill isn't noticeably more painful after plugging in the new TV. We have accumulated our energy habits over time without making the connection.

It could be said that the efficient electronics and appliances available today have made energy cheap again, whether they're saving electricity, gas, or oil. It doesn't cost nearly as much as it did twenty years ago to run a refrigerator or a furnace. However, our monthly bills have not necessarily gone down, at least for long. The technology marketplace has continued to fill our lives with things that require energy to function. For example, there were no personal computers in the seventies, and a lot less electronic entertainment equipment. Prices for electronics continue to drop, allowing the average household to own several televisions or computers instead of just one. The most dramatic example of increased energy use is central air conditioning. According to the U.S. Department of Energy, the percentage of homes with central air conditioning increased from 27 percent in 1980 to 55 percent in 2001, and this percentage continues to rise. So, any progress we made from dramatic gains in technical efficiency is being wiped out by the fact that we are plugging in a lot more stuff. It's no wonder that energy experts are beginning to talk about conservation once again.

Cheap Energy and Time

It's true that all those electric appliances and internal combustion engines have relieved us of the unrelenting drudgery that used to be necessary just to meet our daily physical needs. This is a good thing. But what happened to all that leisure time promised by the early promoters of electric power, or the freedom that owning an automobile was supposed to bring? Instead of using the additional time and flexibility for conversation, recreation, or educational pursuits, we have instead driven ourselves to higher levels of productivity both at home and at work in a mad whirl of multitasking. It could be argued that this choice is also connected to the American ideal of working hard to get ahead, but whatever the reason for this path, we have created a society that depends on the convenience and availability of cheap energy to cram into our days everything that we think needs to be done.

The two-salary household would not exist without washing machines and dryers, dishwashers, microwave ovens, power mowers and snow blowers, hair dryers, supermarkets, fast food restaurants, internet shopping, or a full compliment of vehicles. How else could care and maintenance of the family take place within the twenty-four hours in a day when the two primary adults both have full-time jobs? No wonder we're getting nervous about where our electricity will be coming from or what we will put in our gas tanks. Contrast this to the opposite end of the energy use spectrum where a woman in Africa might spend several hours every day looking for firewood to cook the family's evening meal. Her productivity would markedly improve if she had a solar oven. She might have time to plant and tend another half-acre of crops, or to make some baskets to sell. However, the productivity from her multitasking would be sustainable because she would not be relying on nonrenewable energy to increase it.

Most of us feel occasional frustration about keeping up with our stuff, but recognize that opting out of computers, cell phones, big box stores, and driving everywhere requires serious commitment, not unlike paddling upstream. Even those of us who are sold on the idea of leading a simpler lifestyle have to admit that simplicity is hard. It certainly takes more time, particularly since it's presently so inconvenient. Without cheap energy, life undoubtedly will take more time, and frequently more effort. Breaking the habit of convenience takes incredible commitment.

Of course, if we do nothing, life will likely become more inconvenient, more expensive, and probably a lot less pleasant as we begin experiencing the vicissitudes of global climate change.

Unfortunately, energy efficiency alone will not lower our consumption enough to make coal-fired power plants go away. There is a direct connection between the new coal-fired plant and that old second refrigerator full of beer in the basement. If we really want to scale back on energy use, improving technical efficiencies like miles per gallon or watts per light bulb is essential, but so is choosing to use less. That is the biggest lesson we've learned since energy efficiency arrived in the seventies. The ultimate goal of more effective energy use is lowering overall consumption (and consequently the carbon footprint), whether in homes, buildings, neighborhoods, or whole towns, to a sustainable level that is also healthy for the planet. In figuring out how to do this, we first look at what our energy needs are, how energy is being used, and where it is being wasted. We can then take steps to consciously reduce energy use without sacrificing productivity or comfort.

As we work toward reducing global carbon emissions, advocates of clean coal technology, nuclear energy, or renewable resources may disagree about what our future energy supply will include, but all agree we can no longer afford to waste energy as we have been doing. The population of the earth has doubled since 1960, but the earth's resources have not. Things like land, timber, fish, fresh water, wildlife habitat, and the cleansing properties of the atmosphere are not as endless as they used to appear. Nor are the fossil fuels we've come to rely upon. The less energy we use, the lower the carbon emissions, and the easier it will be to replace fossil fuels with renewable energy sources.

NOTES

1. ACEEE, *Fact Sheet: Appliance and Equipment Efficiency Standards: One of America's Most Effective Energy-Saving Policies,* downloaded August 15, 2007, from http://www .aceee.org/energy/applstnd.htm.

2. USGBC, *Green Building, USGBC and LEED,* June 2007. Downloaded August 28, 2007, from https://www.usgbc.org/ShowFile.aspx?DocumentID=1991.

3. World Building Council for Sustainable Development, *Energy Efficiency in Buildings: Business Realities and Opportunities,* Summary Report, August 2007. Downloaded

August 29, 2007, from http://www.wbcsd.org/DocRoot/MGZ84Tu645qCHVrHwCtl/EEB
SummaryRepoertFINAL.pdf.

RESOURCES

U.S. Environmental Protection Agency's ENERGY STAR: http://www.energystar.gov/

U.S. Department of Energy, Appliances and Commercial Equipment Program: http://
www.eere.energy.gov/buildings/appliance_standards/

Building Energy Codes

American Society of Heating, Refrigeration and Air Conditioning Engineers (ASHRAE)
An international society of technical professionals, open to membership by any person
associated with heating, ventilation, air conditioning or refrigeration through such disciplines as indoor air quality and energy conservation.
http://www.ashrae.org/technology/page/548

U.S. Department of Energy, Building Energy Codes Program: http://www.energycodes
.gov

Alternative Construction Websites

Here are a few websites (known to the author) for alternative construction businesses
and organizations still designing, building, teaching, and encouraging people to learn
about building their own natural home:

Earthship Biotecture, Taos, New Mexico: http://www.earthship.net/

Cob Cottage Company, Coquille, Oregon: http://www.cobcottage.com/

The Last Straw: The International Journal of Straw Bale and Alternative Construction,
Kingston, New Mexico: http://thelaststraw.org/

Earthwood Building School, West Chazy, New York (cordwood, earth-sheltered, and
post and beam construction): http://www.cordwoodmasonry.com/

Green Building Programs

United States Green Building Council (USGBC)
Learn more about the U.S. Green Building Council and the LEED Rating Systems at:
http://www.usgbc.org/

Austin Green Building Program: http://www.austinenergy.com/index.htm
Green home building in Austin, Texas.

Built Green Colorado: http://www.builtgreen.org/
Green building program in the State of Colorado

Energy Efficiency Information

American Council for an Energy Efficient Economy (ACEEE)
A private nonprofit dedicated to advancing efficient use of energy for a healthier economy and a higher quality environment.
http://www.aceee.org/

U.S. Department of Energy, Energy Efficiency and Renewable Energy (EERE)
This list includes several fact sheets about zero-energy homes.
http://www.eere.energy.gov/buildings/info/publications.html

EERE Building America
A public/private partnership program that promotes efficiency in production housing.
http://www.eere.energy.gov/buildings/building_america/about.html

CHAPTER FOUR:

Renewable Energy

The two keys to a sustainable energy future are energy efficiency and renewable energy. In this context, energy efficiency means using the least amount of energy possible to get the job done whether it is through mechanical efficiency or conscious behavior. Energy efficiency comes first because only after we have pared our energy requirements to a minimum do we reap the maximum economy from renewable energy sources.

When we talk about renewable energy, we refer to energy sources that will always be there, at least as far as we're concerned. The sun, theoretically finite, will shine for an estimated five billion years, making its energy infinite from a practical standpoint. As long as there is sun, we will have wind and other energy forces from the temperature gradients of the earth. We will also have biomass sources, in the form of trees, plants, and creatures. These will be available to us as long as our use does not outpace their rate of regeneration. All forms of energy are really solar energy, except for nuclear energy from uranium or geothermal energy, which comes from the earth's core.

Renewable energy comes in many forms. The wind blows because of temperature differences among masses of air and varying topography, primarily created by the sun's heat. Wood and other biomass energy crops like corn and sugar cane collect the sun's energy as they grow. Fossil fuels are merely forms of ancient biomass energy compressed over millions of years into purer, more highly concentrated forms of carbon. The rivers that power hydroelectric dams flow because of the continuous watershed cycle of evaporation and rain or snow, a process in which the sun's heat plays a major role. And, of course we feed our bodies energy in the form of plants and animal proteins that rely on the sun for their existence.

What will we use after fossil fuels? Logic says it will have to be something we will continue to have enough of, and it will have to be environmentally friendly. In other words, it will need to be renewable energy. We have plenty of sun, wind, and organic matter. There is lots of hydrogen, as well as geothermal heat from inside the earth. There is wave power and river power. A lot of research has been conducted and many calculations made over the last forty years about the potential of renewable energy sources to provide for our needs in the future. There is no question that theoretically they can more than do the job.

Talking about the technical potential of these energy sources is a great way to put the concept across, but the bridge between theory and application remains to be built. Despite the abundance of available options, it is not simply a matter of substituting renewable energy for fossil fuels as all the current hype might suggest. There are essential differences between fossil fuels and renewable energy that prevent a simple substitution. For the latter to replace the former, we need an entirely new approach to our energy infrastructure. If we want to make the most of our renewable resources the centralized approach of the current fossil fuel industry won't work.

Fossil fuels are highly concentrated and versatile forms of stored energy that can be easily transported to where they are needed. For example, coal has been (and continues to be) used for heating, transportation, and electricity generation. Natural gas and petroleum are versatile as well, and they can be piped long distances. Consequently, the fossil fuel infrastructure has evolved into a capital-intensive network of pipelines, rail lines, and transmission lines, with immense centralized power plants, refineries, and mining operations.

On the other hand, renewable energy sources are many and varied, and not necessarily interchangeable or moveable. This energy is most efficiently used close by and can be successfully employed at a small scale to provide site-specific energy. Its sheer variety of forms makes renewable energy's role as a magic energy bullet problematic. For those who favor investing massive amounts of capital into one advanced technology to meet the energy needs of the future, renewable energy does not look promising. These are the people who want to take another look at nuclear energy or see us focusing on a "hydrogen economy." But they are looking to continue the centralized, fossil fuel–style infrastructure we currently have, thereby missing the potential opportunities that the many and varied forms of renewable energy represent—opportunities, by the way, that are very much in harmony with the localized economic models currently employed in most sustainable community planning.

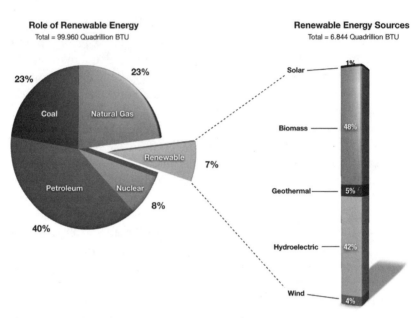

Fig. 14. The Role of Renewable Energy Consumption in the Nation's Energy Supply, 2006 U.S. Department of Energy, Energy Information Administration

Economic flexibility is one example. Renewable energy resources are far more appropriate for small-scale, community-level development than coal or oil. Diversity of energy sources spreads risk and offers economic opportunity for investors at all levels. Our current energy infrastructure employs centralized control of production and transport of electricity, petroleum, natural gas, and coal. Efficiencies of scale generate few new jobs from these sectors, and the immense capital investment required means that ownership and profit go to a very limited number of people, usually not within the local community.

When we conceive of a sustainable energy future, we think about renewable energy as offering an environmentally friendly alternative to fossil fuels because it's cleaner. However, from the broader perspective of sustainable community development, renewable energy offers us the opportunity to spread the wealth created by our demand for energy. Because of the essential qualities of variety and locality, renewable energy sources are a poor fit for the current centralized structure, but ideal for a local and diversified energy economy.

Federal Government Support

The American economy continues to be heavily invested in retaining the fossil fuel infrastructure built primarily in the late nineteenth and early twentieth centuries. Federal government policy, particularly since World War II, has continued to support fossil energy sources whether the administration in power was led by the Democrats or the Republicans.

This is not to say that federal government agencies have ignored the development of renewable energy. The National Aeronautics and Space Administration (NASA) was responsible for much of the early research on solar electric cells, which they used to power satellites. Since 1977, the U.S. Department of Energy has conducted low-key but continued research efforts on a variety of technologies at the National Renewable Energy Laboratory in Golden, Colorado, and at other national labs. The Pentagon has been a significant purchaser of renewable energy technologies, primarily for remote installations.

Most recently, federal support has focused on reducing dependence on foreign oil supplies through research and development of biomass energy sources, particularly as replacements for transportation fuels and other products currently made from petroleum. Ethanol and biodiesel from corn and other plant matter have caught the American imagination as a potential solution to our looming transportation fuel shortages. It could be said, however, that federal support for renewable energy has been doled out in response to specific political energy issues rather than thrown behind a comprehensive clean energy policy for the future.

On the international energy scene, clean energy is considered a major growth industry for the future. The European Union, China, India, and Japan have committed themselves to reducing greenhouse gas emissions, as well as to planning for the decline of fossil fuel supplies. Renewable energy is a primary strategy for accomplishing this. The development of solar, wind, and biomass energies will be shifting the economic and technological order in the world for decades to come. Internationally, we are seeing exponential growth around production of wind and solar electricity. The major international producers of utility scale wind turbines are building new plants to meet the demand, some of which are being located in the United States. Ironically, it was the United States where many of these wind and solar technologies were first developed.

While they can theoretically replace coal, oil, and natural gas for all common energy uses, the various sources of renewable energy are not necessarily interchangeable. In current public policy discourse about energy resources, renewable energy is often spoken about as though it were one thing. "We need to be using more renewable energy," politicians will say, as though it were a single commodity, with interchangeable uses. In their desire to appear proactive on energy issues, national politicians frequently speak of America's dependence on foreign oil in the same breath as wind and solar power, suggesting that, as renewable energy sources, these can solve the problem. Solar and wind technologies don't address transportation at all (unless one recharges an electric car with electricity from the wind or sun).

Even if every home and shopping center in the country were covered with solar electric and solar thermal panels, and every breezy ridge were lined with huge wind turbines, we would still be overly dependent on foreign oil for our transportation fuel, although we would have far smaller coal and natural gas industries for producing electricity and heat. The federal government supports a renewable substitute for gasoline because our oil reserves are essentially gone and we must buy most of our petroleum from other nations. On the other hand, it plays down the potential of solar and wind energy because almost an equal amount of fossil fuel is used to generate our electricity—mostly coal, a domestic energy resource with politically powerful investors.

State and local public officials all over the country are beginning to inventory their solar, wind, and biomass resources and are looking at ways to come out ahead as the clean energy economy emerges. Local and state governments are leading U.S. efforts to reduce greenhouse gas emissions and move to renewable sources of energy. State agriculture agencies and university research programs are exploring local and regional clean energy potential.

Major renewable resources in some areas are readily identifiable. The Southwest will be generating solar thermal electricity, while the Northwest will tap its extensive geothermal resources for both heat and power. The Great Plains states are often referred to as the "Saudi Arabia of wind power," and agricultural states in the Midwest anticipate a high demand for their biomass resources as crops grown for fuel or utilization of farming and forestry wastes. Every state in the Union has at least one renewable energy source it can tap, and some have several. Among the wide variety of biomass resources, localities will claim their own unique applications. In Wisconsin, the methane will come from cow manure. In Iowa, the manure will come from hogs.

Farmers will play a huge role in the development of wind and biomass energy, and prosperity in rural communities will increase as residents become investors in the energy infrastructure. Anaerobic digestion and wind energy are creating new opportunities for farmers, and for keeping farm towns alive. The production of small-scale energy equipment requires new factories that can be located in small towns, creating new jobs and healthier local economies.

Most modern renewable energy technologies are relatively new. While we have already commercialized reliable and efficient wind turbines, solar panels, and anaerobic digestion systems, technological refinements will continue and innovations will emerge. Other electricity-generating technologies, such as solar thermal concentrators or fuel cells have only begun to find successful commercial niches, mostly because they are still experimental and expensive. Clean energy is an exciting idea, and there are many possibilities. Inventors and engineers continue to imagine and innovate, finding ways to extract energy from waste wood and corn sugar. Energy is all around us for the taking. However, our actually being able to make use of all these new ideas is not as straightforward a process as their technological development.

Much work remains to be done to introduce new and different forms of energy production to a population raised on and dependent upon abundant fossil fuels. It may seem comparatively easy to invent a renewable energy technology when you look at what it takes to convince a banker to invest in a manufacturing facility or a utility to hook it up to the grid. Siting solar electric panels or wind turbines frequently challenges neighborhood association covenants or zoning restrictions. The environmental advantages of renewable energy are often overlooked. The cost of coal-fired electricity is artificially low because the environmental costs are not included in the price. Clean energy sources are consequently dismissed by many as being too expensive. While caution and thorough analysis certainly have their place in adopting new technologies, it should be noted that much of the resistance to renewable energy comes from the fact that the United States is heavily invested in fossil fuels and is loathe to change.

There is no question that renewable energy works. The boom in wind and solar power in Europe, Japan, and China certainly makes that point.

Renewable energy technologies and their applications will continue to evolve. Eventually we could see an energy infrastructure responsive to our needs that has minimal impact on the environment and provides a diversified, locally oriented energy economy. We will get there all the faster if we are willing to believe that there is life after fossil fuels.

Renewable Energy Technologies

Before fossil fuels achieved such prominence, we relied on the sun, the wind, and oceans and rivers for our energy, along with wood, peat, buffalo chips, and other biomass that could be burned. As we developed coal, oil, and petroleum, we invented modern technology—engines and turbines, motors, pumps, hydraulics, pneumatics, and electronics, to name a few. It's not surprising that when we started using the nonfossil fuel sources again, we approached them in a high-tech way as well. Every modern renewable energy technology has its own origins, advocates, successes, and to some extent, its own political baggage. What follows are the stories of the current popular renewable energy favorites. Of course, there is no telling what we will squeeze the energy out of next.

Solar Energy

What we think of as solar energy technologies include those that generate electricity and those that concentrate the sun's heat into a useful form. There are even a couple of systems that do both. There is also a distinction between using the light and the heat from the sun. The following are brief descriptions of standard solar technologies.

Solar Electric (Photovoltaic) Cells

There is something almost magical about solar electric or photovoltaic cells (also known as PV), because they generate electricity directly from light. This process, called the photoelectric effect, happens at the atomic level, where photons of light are converted into electrons by the properties of the silicon used to make the cells. The movement of the electrons between materials creates an electric current.

Solar electric panels have no moving parts and therefore operate silently and potentially indefinitely. The electricity they generate is DC

Fig. 15. Solar Flags in Northern Wisconsin. Photovoltaic panels mounted on racks and installed at ground level can serve a decorative function. The rack in the rear is mounted with a two-way tracker that follows the sun up and down during the seasons and across the sky during the day. These solar flags were installed in Sturgeon Bay, Wisconsin, by Lake Michigan Wind and Sun, Ltd. Reproduced with permission from WisconSUN

(Direct Current), which means that solar systems require an inverter to change the current to AC (Alternating Current), the form of electricity used in the United States.

Scientists have known about the photoelectric effect since the early nineteenth-century, but it wasn't actually developed until Bell Laboratories built a solar cell in 1954. Considered an expensive curiosity at the time, it turned out to be an extremely practical source of power for satellites in space. While solar electric systems are still considered expensive, the price tag on the first cells was about $1,500 per watt, compared to around $6 per watt in 2006.

Like many other products developed in the sixties for the space race, solar cells incubated in the NASA laboratories, growing more reliable and less costly over time. By the 1970s, a substantially cheaper solar cell was developed (about $20 per watt) that provided comparatively cheap power for specialized applications such as warning lights on remote towers, isolated communication equipment, and irrigation water pumping. Solar electric panels also emerged as an alternative to fossil fuel generation as part of the national grass roots response to the energy crisis.

In the 1980s and 1990s, solar electricity, which had been developed primarily in the United States, began to capture the attention of the energy industry internationally. Electric rates are much higher in many other countries, as is their commitment to reduction of greenhouse gas emissions, and with increasing demand for solar electricity, a whole industry is rapidly emerging. Production of solar (and wind) generation equipment will continue to grow by double digit percentages in the next few years. Demand has also increased development of production methods that use less silicon in making solar cells, including thin films that can be applied to building materials, roofing, and window glass.

Current trends in Europe and Japan are toward grid-connected, customer-sited systems, with development of utility-scale installations. The international photovoltaics industry is growing by 15 to 20 percent per year. During the second half of the last decade, worldwide shipments of PV panels tripled. In the United States, where PV has mostly been associated with remote, self-sufficient homes, grid-connected solar electric systems are gaining popularity. The State of California leads in its support of grid-connected systems and other solar technologies. The establishment of Renewable Energy Portfolio Standards in many states has encouraged utility programs to promote grid-tied systems.

Advantages

Considering the fact that sunshine is pretty ubiquitous, a nonmechanical and expandable technology for collecting its energy possesses a number of distinct advantages. Because PV panels are modular, the design of a system can respond to site-specific requirements, and can quickly be expanded or reduced as needs require. You can hook up one panel or ten, depending on what you need.

PV systems can easily be made portable and they require only sunlight to start working. PV enjoys great popularity as a way to provide electric power for recreational vehicles and portable electronic equipment. Also, trailer-mounted PV systems are frequently used to provide emergency power in the wake of hurricanes, earthquakes, or other disasters. PV systems work anywhere the sun shines. In areas where clouds are common, a larger system is required to get the same amount of power available in sunnier locations. Sunlight reflected off snow increases PV panel production, and panels operate more efficiently in cooler temperatures. Also, PV systems produce the most electricity when the sun is at its zenith, typically when the air conditioning load is at its peak in the summer. Businesses can use grid-connected PV to avoid higher utility costs for peak electricity, and residential grid-connected systems can contribute power while their owners are away at work.

Photovoltaic cells are similar to electronic components because they are solid state in nature and are therefore extremely reliable. Theoretically, PV panels could last at least a lifetime. Warranties on PV systems vary, but a twenty-five-year warranty from the manufacturer is not uncommon.

Solar electricity has earned its green reputation for good reasons. First, photovoltaic technology generates electricity from sunlight, an energy resource that is continuously delivered and requires no refining process to be useable. Nor does this technology produce greenhouse gasses or particulate air pollution. Using solar electricity means offsetting the use of fossil fuel or nuclear power generation, thereby reducing the emissions that contribute to global warming, and other pollution or security problems. Producing 1,000 kWh of electricity with solar power reduces sulfur dioxide (SO_2) emissions by nearly eight pounds, reduces nitrogen oxide (NO_x) emissions by five pounds, and reduces carbon dioxide (CO_2) emissions by more than 1,400 pounds. Closer to home, PV systems produce no noise pollution because they generate electricity silently.

Many people wonder about the environmental impact of manufacturing the system components themselves. The primary material in the cells is silicon, a nontoxic chemical element also used for making glass, ceramics, and semiconductors. It is one of the most common elements on earth. According to a recent report funded by the European Commission, the amount of energy used in producing the cells and other components, including the panel mounting frames, will take the panel under two and a half years to generate once it is installed. This is a considerable improvement over manufacturing methods in the 1970s, when questions about the environmental friendliness of panel fabrication were first addressed. Efforts continue toward improving the efficiency of manufacturing solar electric components. The primary goal is reducing costs, but the industry is aware of the environmental imperatives as well.

Disadvantages of PV

The primary disadvantage of PV systems remains their cost, because manufacturing PV panels is expensive. Solar electricity will become economical as the cost of other forms of generation increases and new technological developments for manufacturing panels reduce costs per unit. The cost of solar electricity is customarily compared to the going rate for electricity from the local utility, which is usually from coal-fired power. Utility rates in most of the United States make solar electricity look pretty expensive. In other countries, however, where governments take account of the environmental impacts of fossil fuel generation or must import their fuels, solar-generated electricity is far more economical.

Because of the comparatively high system cost in the United States, not all geographic areas are ideally suited for solar electric installations. Areas with cloudy skies or an abundance of shade trees or tall buildings are at a disadvantage. As solar electricity becomes more economical, however, geography will play a smaller role. The large, flat roofs of big box retailers and shopping malls anywhere in the United States will provide excellent, nonshaded opportunities to generate solar electricity.

One last concern is the amount of space needed for a solar system at current panel efficiencies, which equal 12 to 15 percent on average (meaning the maximum amount of available solar energy they are able to capture per square meter of sunshine). There are a number of impacts on system performance including latitude, shading and cloud cover, and whether the panels are tracking or fixed. However, a very rough estimate of the amount of space needed for a 1-kilowatt system is about 100 square feet. A 1-kilowatt system can produce between 1,100 and 1,500 kilowatt hours of electricity per year depending on latitude and other conditions. The average household usage is between 10,000 and 12,000 kilowatt hours per year. Using these assumptions, a system in a northern latitude location would need to be 9 or 10 kilowatts in size and occupy 900 to 1,000 square feet to supply the needed electricity. Using $6,000 per kilowatt as an assumed installation cost, this system price would be $54,000 to $60,000. However, many households could cut their electricity use by half or more without feeling a pinch in convenience. It is easy to see why cutting energy use through efficiency can substantially reduce the cost for a solar PV system.

Solar Thermal Generation Technologies

It is also possible to generate electricity using the sun's heat, but the process requires heating a liquid to generate steam that runs a turbine, so it is called solar thermal generation. These systems really do need to be located in areas where sunlight is almost constant year-round.

Over the years, the federal government has focused research efforts on solar concentrators because they have the potential to produce solar electricity at a greater efficiency than PV and are therefore of greater interest to utilities for generating large amounts of renewable power. Several types have been explored at Sandia National Laboratory in Albuquerque, New Mexico, Lawrence Livermore Laboratory in California, and the National Renewable Energy Laboratory in Golden, Colorado, all locations with sufficient sunlight to support the technology.

Trough Concentrators

A curved, reflective trough concentrates solar heat on pipes suspended lengthwise along long rows of troughs. Liquid (a type of oil) running inside the pipes is heated to around 300 degrees Celsius. This liquid is collected in a boiler where the heat boils water to generate steam to run a turbine. Trough concentrators are the simplest solar thermal generation technology and have been used in California since 1985 to produce about 350 megawatts of power for California Edison in nine separate installations in the Mojave Desert.

Parabolic Dish Concentrators

Theoretically more efficient than parabolic troughs, dish concentrators track the sun as it crosses the sky. Mirrors are positioned in a dish-shaped configuration of concentric rings focusing reflected heat on a central point, achieving much higher temperatures. The complexity of these systems has been problematic in using them in commercial utility installations. The most successful type of dish concentrator uses a Stirling engine as the central power generator, which operates from the liquid sodium heated by the concentrated sunlight. In 2005, California Edison signed a power purchase agreement for 500 megawatts of dish Stirling power, to come from a 4,500-acre installation in the Mojave Desert. This installation will eventually be expanded to 850 megawatts and will be the largest solar generating facility in the world.

Central Receiver or Power Tower

These massive systems represent the largest and most complex solar generation technology developed so far. First begun in the 1980s, they represent what can be achieved by the deep pockets of the federal government when it is politically motivated to generate utility-scale solar electricity. Central receiver plants based on the experimental U.S. installations are being developed in Spain and South Africa.

The operating principle is using a field of mirrors called heliostats facing south, but adjusted to concentrate reflected sunlight toward a tiny area at the top of a 300-foot tower. The technology is based on the simple idea of the magnifying glass that sets a pile of dry leaves alight by concentrating the sunlight. The intense heat is used to boil water to generate steam for electricity generation. Refinements in the design involve the use of molten salt as a heat storage medium, allowing electricity generation after the sun has gone down. Experimental models were built by the U.S. Department of Energy in the Mojave Desert and New Mexico, but the only commercial power tower installation currently existing is the PS10 Solar Power Tower in Seville, Spain. This tower uses 624 heliostats aimed at the 377-foot tower to produce eleven megawatts of power.

Solar Water Heating (Solar Thermal Systems)

Anyone who has felt the warm water coming from a garden hose sitting in the sun understands the principle of a solar thermal water heater. The two primary challenges to practical application of this simple idea are first to achieve maximum efficiency, or heat as much water as possible while the sun is available, and second, to keep the water hot after the sun goes down.

In many ways, hot water helps make our lives comfortable and healthy. We use it to keep ourselves, our clothes, and our homes clean.

Fig. 16. A Solar Home. This home in northern Wisconsin has two solar systems. On the left are flat-plate solar hot water panels, and the panels on the right generate electricity. Photo by Bob Ramlow, reproduced with permission.

It's hard to imagine life without it, but hot water available on tap is a relatively recent phenomenon. For millennia, people sought out hot springs for bathing and washing or heated water in a pot over a fire. More elaborate stoves incorporated a water tank next to the firebox that kept water warm, using wood or coal for fuel. Coal oil water heaters were used in urban homes in the nineteenth century. The modern water heater as we know it, which heats and stores hot water to be available on demand, was invented by Edwin Ruud in 1889.

In rural and frontier America, however, these modern conveniences were frequently unavailable, and where fuel was scarce, ingenious settlers turned to another source of heat. The first solar water heaters were metal tanks painted black that would soak up the sun's heat during the day, providing hot water in the early evening, but they would lose most of their warmth as the sun set. In 1891, Clarence Kemp of Baltimore patented the Climax Solar Water Heater, an invention that addressed the heat loss problem by encasing the black metal container in a glass-

Fig. 17. Advertisement for a Climax Solar Water Heater Graphic courtesy of John Perlin/ Ken Butti

covered box. The Climax had mild success on the East Coast but took California by storm in the first decade of the twentieth century.

Eventually, the Climax was overtaken by a more advanced design that separated the solar panel from the indoor storage unit, allowing the heat to be retained longer. By 1918, solar energy was the cheapest and best way to heat water in California, and thousands of systems were installed. The industry crashed, however, with discovery of natural gas near Los Angeles in the 1920s and 1930s.

Solar water heaters migrated to Florida when growth in home building exploded in the 1920s. With space heating being essentially unnecessary and fuel very expensive, the success of the solar technology proven in California was a good fit. By the early 1940s, half the homes in Florida had solar hot water heating systems. Decline of the technology in Florida, however, was not discovery of local natural gas supplies but rather a greater availability of cheap electricity after World War II.

During the oil embargo of the 1970s, the United States experienced a major shock to its independent self-image. The feeling of vulnerability was uncomfortable, and an energy panic entered the political consciousness that produced both positive and negative effects on the country's energy culture. On one hand, this was a time of great innovation and inspiration, as people from the federal government to the grass roots started taking the potential for renewable energy seriously, particularly solar technologies. It is considered by many to be the time when the modern age of renewable energy began. However, some good intentions of that era proved counterproductive. In 1979, generous tax credits offered by federal and some state government programs inspired a market demand for solar water heating systems so tremendous that the fledgling industry was unable to consistently deliver the quality and reliability a more gradual market development would have allowed. Hundreds of small manufacturing and installation companies started up, many without adequate experience or skill. Some were idealistic; others were simply in it for the money.

When President Ronald Reagan abruptly canceled solar tax credits in 1986, the solar water heating industry not only fell apart, it left behind a reputation for leaking roofs, shoddy installations, and what became known as "orphan systems," solar water heaters that didn't work and couldn't be fixed because the equipment dealer no longer existed. For the next decade, the general public perception of solar water heating was that it might even be a scam, but it was definitely something that didn't work and couldn't be trusted. This perception still clings in many minds.

The considerably downsized solar water heating industry soldiered on, developing reliable systems that work in cold climates, and improving and refining installation techniques and standards. Today the North American Board of Certified Energy Practitioners (NABCEP) offers a rigorous national certification program for solar water heater installers that compares to certification in other building trades. The Solar Rating and Certification Corporation (SRCC) is a nationally recognized independent testing facility that certifies solar thermal panels based on a consistent standard. The industry has come a long way toward living down its carpetbagger reputation from the seventies.

Even though solar water heating technology and the reputation of its industry have both improved since the tax credit era, its popularity continues to depend on what the public perceives as economical. However, as we face potentially declining supplies of natural gas and rising electric rates, and the threat of global warming is giving us pause about burning fossil fuels, solar water heaters are starting to look pretty good again.

There are several types of solar water heaters suited for different climates and for producing different ranges of temperature. The least expensive solar water heaters are those designed for warm climates or warm climate conditions where freeze protection is unnecessary. These include the heaters used even in cold climates for seasonal outdoor pools.

Protection against freezing pipes is the trick for systems used year-round in cold climates. These systems use a heat exchanger in the storage tank and antifreeze circulates in a closed loop pipe between the solar collector, typically the flat-plate type, and the tank. The evacuated tube collector is the most efficient type and also the most expensive, used primarily where high temperature water is needed, usually for space heating systems. The collector pipes are encased in glass vacuum tubes that reduce the heat loss.

Solar Water Heating Applications

The current number one use of solar energy in the United States is for heating swimming pools. In warm weather states, pools can be heated year-round with thermoplastic pool heating panels, as can seasonal pools in colder areas. The cheapest pool heater to install uses natural gas or propane, but the fuel is expensive and getting more so. A solar pool heater will pay itself back in two to four years from the energy savings. A solar water heating system for domestic hot water is perhaps the next best renewable energy investment for homeowners with a sunny, south-facing roof. Solar heated water can also be used for space heating in under-floor piping systems or radiators. When a solar water heating system provides less than 100 percent of hot water requirements, it still saves energy by preheating water going to the conventional water heater or heating system boiler. Solar water heaters are proving to work well for multifamily buildings as well as detached homes.

The commercial sector is beginning to recognize the savings available from solar water heating. Commercial systems heat water for restaurants, hotels, car washes, and commercial laundries. Any business that consistently uses large quantities of hot water may find a solar system to be economical. Similarly, municipalities are beginning to use solar hot water systems for hospitals, athletic facilities, pools, correctional facilities, senior living centers, maintenance garages, and fire stations.

Fig. 18. *Solar Thermal Air Heater, Mary Coffrin Hall, University of Wisconsin–Green Bay* Photo reproduced by permission from WisconSUN

Solar Thermal Air Heating

There are also systems that use solar energy to preheat make-up air for high-volume ventilation systems or to provide space heat. They are typically used for warehouses or agricultural buildings, or to provide make-up air for restaurant hoods. These systems draw outside air into a broad and shallow chamber covered with a dark metal screen that is being heated by direct sunlight. Small ventilation fans direct the air into the building through ducts that distribute it. These systems are economical even in cold climates for providing a notable percentage of facility heat that would otherwise be produced by natural gas or propane.

Passive Solar Design

A cat will choose to sleep in a patch of sunshine on a rug simply because it's the warmest spot in the room. There's no technology involved, just the fact that the sun brings both heat and light through the window. Passive solar design seeks ways to take advantage of this simple principle.

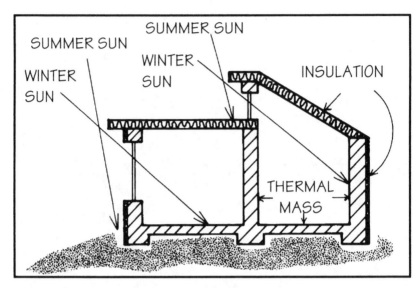

Fig. 19. Winter and Summer Sun Angles Illustration by the author

First, the building must be oriented so that a majority of the windows face south, and the site must allow the sun access without excessive shade. The design must allow entrance to the low winter sun but block the high, hot summer sun. This is typically accomplished with an overhang, calculated for the angle of the sun in that location.

Once the sun's heat enters the space, there must be a way to collect and store it, or the space will simply be overheated. The building design must therefore include materials that provide thermal mass inside the house to soak up the solar heat during the day and store it for slow release at night. Examples include stone, concrete, adobe, or brick. These can be used for floors, walls, or fireplaces.

Efficient construction of the building shell, including high levels of insulation, is extremely important, because the idea is to capture as much of the sun's heat as possible and hold on to it. Passive solar design can provide all or most of the heat for a house in sunny climates where winters are relatively mild. In northern climates passive solar heating can easily provide 25 percent of the home's heating requirements. In these colder climates, passive solar designers place greater emphasis on the concept of daylighting rather than heat as a primary energy-saving strategy. Where it's cloudier, daylight is welcome both for the cheerful atmosphere it creates and for the opportunity to leave the lights off to save electricity.

Other passive solar design strategies include sun spaces and attached greenhouses that can essentially create a buffer zone in the winter and an additional microclimate for growing things. During the transitional times of spring and fall a sunspace or greenhouse becomes an outdoor room that extends the warm season. Wherever a passive solar house is built, it will produce the highest level of comfort both winter and summer with manual operation of insulated curtains and windows for cross-ventilation. It is a house to live with, rather than simply to live in.

In its Leadership in Energy and Environmental Design (LEED®) Certification process, the United States Green Building Council awards a design point for daylighting 75 percent of spaces in a new commercial building to a prescribed level. Daylighting is seen as integrating both energy savings and work environment quality into the building.

Wind Power

Wind has been sailing boats, grinding grain, and pumping water for millennia. Electricity generation was added to the list in 1920s rural America, as farm families yearned for the electricity made available to city residents. The market boomed for small one- to three-kilowatt wind generators, employed to charge up batteries for radios and lights in farmhouses. At this time, several hundred thousand small wind power generators were operating on farms both in the United States and abroad. When grid electricity arrived in rural America by way of the Rural Electrification Administration, the market for these small machines declined, disappearing almost completely by 1950.

It was the energy crisis of the 1970s that saw the revival of wind-generated electricity and the beginning of what has become the fastest growing renewable energy technology today. The first utility-scale wind farms were developed in the 1980s, spurred by the Public Utility Regulatory Policies Act of 1978 (PURPA), which made it possible for nonutility generation developers to sell their power.

However, the grassroots wind power pioneers in the 1970s were not thinking about wind farms or even selling their electricity. Their motivation was political, born of other events in the 1960s. They were alarmed by the oil embargo and disturbed by the environmental impacts

Fig. 20. 660-Kilowatt Wind Turbine in Byron, Wisconsin, owned by We Energies, Milwaukee
Photo by the author

of coal-fired generation. But these concerns were seen rather as symptoms of a much larger issue. When soldiers returned from World War II in 1945 to build a good life for their families, the United States had just successfully won a war. The Viet Nam War ended less decisively in 1975 after a decade of dissension at home. Weary troops returned to a nation still bitterly divided, many mistrustful of the federal government's motives and priorities. Many of those who served in the war as well as those who protested it felt they had seen the military-industrial complex up close. The life they wanted to build for their families would be a good one, but they would build it with their own hands on the land, as far as possible from the false promises of centralized, corporate, or government control.

Energy was a strong symbol of dependence on the central power structure, particularly in light of the oil shocks of 1973 and 1979. Generating one's own electricity off-grid was a gesture that simultaneously defied and improved upon the old order. Wind turbines, perhaps more than the high-tech and more experimental PV panels, became symbolic of deep-seated rebellion against mainstream American life. Public attitudes toward wind and solar power, and indeed toward evidence of a connection between global warming and fossil fuel combustion, are still colored by the political atmosphere of the seventies when alternative energy and environmental awareness equated with protesting America's mainstream values.

When utility-scale wind power began its growth in the 1980s, a different group of advocates and developers emerged. They also believed in the technology, but they envisioned it harnessed in large, centralized "farms" of wind turbines, producing electricity for the utility grid. However, both the grassroots wind advocates and the utility-scale developers recognize that lack of federal policy to promote clean energy sources affects all efforts to move it forward, whether those efforts are in behalf of individual energy independence or development of centralized grid energy. They are therefore, finding ways to work together while each maintains its own perspective on the future of wind energy.

As the American wind industry matures, a new perspective is emerging in the form of community wind development. These are turbine installations of one or a few utility-scale turbines owned by local, independent corporations or cooperatives, which sell their power to a utility. These turbines are generally smaller utility-scale machines that can be located to take advantage of very local wind conditions that would not support a large wind farm. Municipalities, farm coops, and private investor groups, working with progressive venture capitalists and banks, are beginning to develop local investment opportunities in wind generation. This is not unlike the Danish model, where many individuals own shares in wind farms.

Political and economic progress, however, has been rocky. Wind energy development is now a fluid and international phenomenon among the players. Vestas of Denmark, Enercon of Germany, and Gamesa of Spain are three major companies in the international utility-scale turbine market, reflecting the political commitment to wind energy of their respective governments. India and China are also ramping up their manufacturing capability, both for domestic and international projects. Wind turbine manufacturing in the United States has been hindered by lack of federal political support for long-term renewal of the production tax credit for wind energy, which was first introduced in the Energy Policy Act of 1992. This tax incentive was created for wind power only and is similar to those offered for fossil fuels. Its purpose is to provide the market stability necessary for major capital investment in turbine manufacturing facilities. So far, Congress has only renewed it for two or three years at a time, making the long-term investment in turbine manufacturing facilities for U.S. wind projects a risky venture.

The Wind Resource

Obviously, wind turbines must be located where strong winds are blowing. The national Renewable Energy Laboratory Wind Resource Map (see resources at the end of this chapter) shows fair to excellent wind resources from the Dakotas south to Texas, with large areas of good to excellent wind resource areas scattered throughout the West. Consistently windy conditions occur in hilly and mountainous terrain as well as over open water. This can be seen on the map, where good conditions are identified along the Appalachians and over the Great Lakes. The southeastern part of the United States is conspicuous for its lack of viable wind resources. The wind map does not tell the whole story, however. Of the two states presently generating the most wind power, the American Wind Energy Association ranks Texas as second in the twenty windiest states (behind first-place ranked North Dakota), with California ranked only seventeenth.

Discounting the political and economic factors surrounding the electricity industries from state to state, there are several factors that influence whether or not wind turbines are sited and where this might occur. National and State wind resource maps only give a general idea of where the resource is best. Specific local conditions must be measured, usually over a period of a year or more. Assessment of a potential site will include average wind speeds and proximity of structures, trees, or geographic features that would cause wind turbulence and reduce the efficiency of turbines.

Turbines work best in open areas away from population centers, and away from the potential interference with airport runway paths. For this reason, wind farms have been most often located either on ridges or hilltops, or in broad, open agricultural areas where there are no forests or structures to interfere with the wind flow. Wind energy is currently seen as a rural energy source, primarily because of the need to avoid turbulence and because of the total amount of open area a wind farm requires.

Wind Turbine Types

The familiar three-blade design is known as a horizontal axis wind turbine, meaning that the blades turn a horizontal shaft that drives the turbine to generate electricity. This is the design that works best in current applications, whether large, utility turbines or small, residential models. Vertical axis wind turbines also exist. The best-known model is the Darrieus machine, which resembles a giant egg beater. These machines are designed to be built close to the ground, eliminating the need for a tower. While a few successful installations of vertical axis machines do exist, the design itself is not as efficient as the propeller-type horizontal axis machine, and wind speeds are usually much lower near the ground, so they have not been cost efficient. There are experiments to install vertical axis wind turbines on buildings in urban areas to harness city wind, but so far no design has been completely successful. One difficulty is that the turning blades cause vibrations that can damage the structure of the building. The dream of turbines atop skyscrapers has yet to be realized.

Turbine Towers

Turbines are installed on towers, which can be one of several types. The most common tower type for large turbines is a tubular steel tower that is manufactured in sections of 20 to 30 meters each and delivered to the site on a flatbed truck. These towers are constructed atop hefty concrete foundations, with the tower sections lifted into place by a crane. As turbines become larger and towers necessarily become taller, the cost of raising the tower increases dramatically, offsetting the efficiency of the larger turbines. Another type in common use is a lattice tower, a type also used for radio and cell towers. They are generally used for smaller, utility-scale turbines

The height of a wind turbine tower relates closely to the turbine's ultimate efficiency. Wind speeds increase exponentially as height is extended. Taller towers lift the turbine away from ground turbulence and into a more consistent wind stream.

Windy NIMBYism

Although wind turbines have emerged as the first major source of renewable electricity in the United States that is close to competing economically with traditional fossil fuels, this new industry is not without controversy. Wind farm developers in the mid-1970s tended to erect their wind farms on remote mountain peaks and passes, but the present crop of wind advocates has begun turning to agricultural lands. Wind turbines are strange new creatures on the horizon, and for a number of reasons they are not always welcome in the neighborhood. Some objections are based on the simple emotional response to the unfamiliar. Other objections are rooted in cultural or economic concerns. Wind farm controversies are popular media fare as well, which can magnify local disagreements and delay resolution.

Organized groups of protestors have halted or delayed wind farm projects in several states. The most famous controversy so far has been the Cape Wind project offshore from Cape Cod. This project is to have 130 turbines for a total rated capacity of 468 megawatts and will produce three-quarters of the electricity needed to power Cape Cod.

Protestors of this project have included both environmental groups and vacation home owners who are collectively concerned about impact on marine species and the visual impact on the horizon. In this instance, many of those objecting are high-profile residents of Cape Cod, which has made this a national news story and detracted from a sound assessment of the project. Similar objections have been raised about other, smaller projects elsewhere, and while the participants may not be celebrities, the emotions run equally high.

For some, lack of familiarity with the technology makes it feel like a threat to their health and safety. Common fears are blade flicker causing epileptic episodes, blades flying off in high winds or flinging ice great distances, current or voltage irregularities that could harm farm animals, or unreasonable noise levels. Once most people have the opportunity to learn how turbines operate or have a chance to visit a wind farm, these are no longer issues. They remain, however, the rural version of urban legends.

There are other fears and concerns less easily dismissed, and these are usually rooted in economics. First is the fear that proximity to a wind farm will reduce a home's property value. This issue has emerged since wind farm development has taken root in populated agricultural areas, as opposed to the relatively remote ridgetops and mountain passes where they were first located. The property value issue is very difficult to pin down, primarily because there are very little reliable data to study. Homes are scattered far and wide in truly rural areas, and so far there have not been enough wind farms built, or properties nearby that change hands before and after the development, to be able to draw meaningful and reproducible conclusions. Another difficulty is determining just what is visually or audibly unpleasant enough to drive down the price of a property. Simply plotting the distance from the nearest turbine doesn't necessarily determine anything, because trees or hilly terrain could hide it from view. There have been several studies done to address this question, but none have been able to identify a direct correlation between turbines and lower property values. On the other hand, there are purported rural housing developments located specifically with views of wind farm turbines as a selling feature.

Perhaps it can be said that property values in the country are defined differently than those in the city. The question of property values in relation to wind turbines is less likely to be of importance in more deeply rural areas, away from the edge of sprawl. Generally speaking, farmers welcome the opportunity to farm the wind that blows across their land. The actual footprint of a turbine is quite small (less than 1 percent of a wind farm's land area), allowing crops and grazing to continue uninterrupted underneath. Whether a farmer is offered a lease by a wind developer in the area, or whether the local farming cooperative finances a community wind turbine of its own, rural communities are beginning to enjoy new financial opportunities from their land.

Most conflicts that arise occur where the city is moving into the country. City folks who buy a place to live in the country, but still work in the city, want their home property to retain its rural charm. They are not invested in the rural economy and are not interested in the advantages that make turbines so attractive to farmers. Wind farm developers have become aware of the difficulties a project can encounter when attempting to locate it near more populated areas. It is common now to offer financial compensation to owners of neighboring properties, as well as payments to local governments to garner support. Wind developers have learned that avoiding general negativity and expensive delay is best accomplished by involving the community early in the project and responding in a timely manner to local concerns, both real and imagined.

Wind Turbine Zoning Ordinances

Communities can be proactive about wind turbines by including a section in their zoning ordinance that permits their installation and outlines restrictions. If a local government opens the discussion among residents before a wind turbine project is proposed, people have the opportunity both to learn about the technology and to express their feelings without having to take sides in a controversy. It is also an opportunity for the local government itself to learn about effective wind ordinances adopted elsewhere that they might use as models.

The ordinance might distinguish between projects that would provide power on site and those that would sell power to the utility grid. Typical restrictions might include maximum tower heights, noise levels, and setbacks from property lines. The ordinance could also include the standards and codes with which the project needs to comply in order to keep its permit.

Communities in geographical areas with a good wind resource will be more likely to need an ordinance that applies to utility-scale wind farms, and they may wish to consider a variety of other restrictions or allowances. Local officials are well advised to learn as much as they can about wind turbines, particularly by talking to communities where wind farms have already been built. The error would be to do nothing until a full-blown confrontation erupts. Wind power development can be of economic benefit to rural communities, providing revenue as well as creating local jobs.

Wind Turbines and the Environment

The visual impact of wind turbines is pleasing to some and ugly to others. It is interesting to note that opinion polls have shown that people get used to seeing wind turbines on the horizon the way they get used to seeing cell towers, transmission lines, and fast food restaurants. When you realize that there are areas in the world where coal-fired power plants and heating systems produce such clouds of pollution that people can't see much of anything at all, wind turbines could be regarded as helping us see everything more clearly.

The most controversial issue regarding the environmental impact of wind turbines is the lingering misperception that the moving blades kill great quantities of birds. This misperception has its roots in the unfortunate siting of an early wind farm. The first large wind farm in the United States was built in Altamont Pass in Northern California in the late 1970s. The geographic configuration of Altamont Pass produces steady winds, perfect for both wind turbines and raptors hunting for food, such as golden eagles, red-tailed hawks, and other protected species. Once the turbines were installed, birds were being caught by turbine blades and electrocuted by exposed wiring as they dove for their dinner. Altamont Pass is also a migratory bird route, which added to the mortality rate at certain times of year. The original designers of this wind farm were not aware of potential environmental impacts when they erected the turbines, and the project's bird mortality has haunted the entire wind industry ever since.

This wind farm is still operating, with around 7,000 wind turbines, mostly smaller than models currently installed at utility wind farm projects. These older models, with rapidly turning blades, are gradually being replaced with the taller and more powerful new turbines that employ a more efficient gear ratio. The blades on these turbines turn much more slowly, making them far less dangerous for passing birds. Efforts continue to make the Altamont Pass wind farm safer for resident birds, but experts concede the mortality rate may never be as low as it is at currently developed sites elsewhere.

This issue has been taken very seriously by environmentalists and scientists and continues to be heavily studied. Recently, turbine impact on bats has been added to the question as well. The legacy of Altamont has made Environmental Impact Statements standard practice for wind farm development, with particular attention to bird migration and nesting patterns. Wind farm designers take bird and wildlife patterns into account as they site the turbines. Potential bird mortality contributed to the general adoption of cylindrical towers over lattice towers, which were known to attract birds for perching and nesting within their structural members. Recent studies have shown that birds are generally unaffected by the larger, slower turbine blades, and as the towers get taller, they tend to exceed the altitude of most bird flight patterns.

The National Wind Coordinating Committee, a collaborative comprising public, private, and nonprofit sector wind energy stakeholders, commissioned a study in 2001 that addresses bird collision mortality from various manmade structures for comparison.[1] Their statistics show the average number of birds killed each year per turbine is 2.9, which includes the higher California averages. Without the California numbers, the average is 1.83 birds killed per turbine. The study estimates a total of 10,000 to 40,000 bird fatalities per year by the 15,000 wind turbines installed in the United States as of 2001. For the 80,000 existing cell towers, it estimates the number of bird fatalities to be four to fifty million birds. This calculates as being between 50 and 625 birds killed annually per cell tower. The category with the highest estimate is windows and buildings, which average 98 million to 980 million total collision deaths per year.

Studies are now focusing on wind turbine impacts on bats, which have their own migratory patterns and feeding habits. The issue was raised after a significant number of dead bats were found around a wind farm in West Virginia. So far studies indicate that bats living near turbines seem to coexist with them, but mortality rises during migration. However, there are widely varying mortality patterns depending on location. Much more study is needed, not only about bat interaction with turbines, but how bat migration patterns are being affected by a whole host of human-caused circumstances.

Biomass Energy Sources

There is a great deal of energy stored in organic matter, both plant and animal. Apart from the plants and animals we eat, wood is the most common biomass energy source we use. We have also burned other organic materials for heat, including peat and dried manure. Modern technology is discovering other, more efficient and flexible ways to harness the energy in biomass. Currently, three areas of great potential are wood waste and methane gas for heating and electricity, and biofuels for transportation. Much of this research and development is very new, and it's difficult to predict what directions development of these potential energy sources may take.

Agriculture and forestry experts are already taking inventory and studying where the biomass will come from: forest residue, manure, crop residue, food waste, fuel crops including corn and sugar cane, trees, grasses, and others. On the other hand, in addition to transportation fuels made from biomass, scientists are working on duplicating the wide variety of other products we now make from petroleum such as plastics, chemicals, lubricants, and solvents. A lot of different industries are eyeing available biomass resources as substitutes for petroleum. As we are forced to eliminate fossil fuels, we will need to coordinate how we use our biomass resources very carefully.

The development of biomass resources has taken place primarily in university agriculture research programs and private corporations. The farm lobby has gathered considerable political support at the national level for ethanol and other biofuels, and utilities in some states have invested in anaerobic digestion demonstration projects on dairy and livestock farms and burning wood waste and other organic matter in combination with coal.

Unlike solar and wind energy, biomass is not a grassroots renewable energy movement. The players are corporations, national engineering firms, large farm operations, state and federal government officials, and venture capitalists. There is a lot of money to be made for those who find a substitute for gasoline (among other applications), so community development and environmental friendliness are sitting firmly in the back seat. This is not to say that smaller and more local operations can't share in the benefits of these new energy opportunities. Biomass resources, like wind and solar, will ultimately deliver the most benefit at the local level because it's too expensive to transport the raw material far.

Wood

Wood has been used as a fuel by humans for thousands of years. For the last two hundred years or so, use of fuel wood declined because easily transportable supplies near settled areas in the United States had been used up. People turned to burning coal in their home fireplaces and stoves because wood became too expensive. Much of our forested land has regenerated since that time, and there are more trees growing now in the longest settled parts of the country then there were at the height of the fuel wood era. Burning is the least valuable use for wood, perhaps the main reason we stopped burning it when supplies got tight. It was much more profitable to build houses with it and to make furniture and paper. Many states still consider their forest resources to be important to the local economy. The forestry and paper industries continue to find ways to increase efficiency. They now return most of the waste generated to the manufacturing process.

There are still waste streams of wood that go unused, such as discarded pallets and packing materials, construction waste, waste from urban forestry, and what is left on the forest floor after logging operations. Except for the latter, this waste usually ends up in landfills. Now that fossil fuel is getting more expensive both economically and environmentally, this material has accrued some value as businesses and municipalities and even utilities are reconsidering wood waste as potential fuel. New wood-burning commercial boilers and furnaces with automatic feed mechanisms make using waste wood chips or pellets for space or process heating almost as convenient, and frequently more cost effective, than natural gas or propane. Such a system is only economical, however, if a reliable, quality source of wood waste is available nearby, usually within a fifty-mile radius. Currently, these systems are most popular with users like wood products manufacturers or sawmills that can burn their own wood waste. They might also make a deal with a nearby company or school to sell them their wood waste for space heating.

Many people in the United States, particularly in remote areas, use wood as their primary home heating source. They continue to get their fuel from local wood lots, often on their own property. The wood stove industry serves these customers, along with many home owners who want energy-efficient fireplace inserts or a wood stove for the great room. As natural gas prices keep rising, even more people are looking at wood as a less expensive alternative for heating their homes, either in wood stoves, furnaces, or efficient fireplaces. Burn efficiency and emissions for these appliances have greatly improved over the years as a result of gradually tightening regulations by the Environmental Protection Agency. All wood stoves manufactured after July 1988 had to emit fewer than 8.5 grams of particulates per hour. In 1990 that was lowered to 7.5, where it remains today. As designs continue to improve, a lower standard of 4.5 grams will likely be instituted. When the EPA began investigating pollution from wood stoves in the 1980s, the typical emission level was 40 to 60 grams per hour.

The latest manifestation of wood-burning technology is the wood pellet stove. The fuel is small wood pellets made primarily of hardwood sawdust waste. The compressed pellets, which resemble rabbit chow, are held together by the natural lignin in the wood. These are so efficient when they burn that there is little smoke or emissions, and the stoves don't require a masonry chimney but can simply be vented through the roof or out a wall. Little ash is produced, depending on the grade of pellets. The stoves do require electricity to work because of the pellet feed and blower mechanisms. Once again, a reliable supply of pellets is necessary locally to make this cost effective. Some of these stoves are designed to burn pellets made of other biomass materials or corn.

The pellet stove is an eco-friendly way to make use of sawdust and other biomass waste materials as combustible fuel. Another recently popularized heating appliance, the outdoor wood furnace, is not. These furnaces are located outdoors in small sheds. The firebox heats a surrounding water jacket, and the heat from the fire is transferred to through water pipes that run to the building being heated. These units have become infamous for producing copious amounts of smoke, and their installation is banned in many local jurisdictions. The concept of such a furnace is sound in theory as a way for rural homeowners to use their own wood resources, but they have a serious design flaw. The water jacket tends to cool the firebox, preventing the fire from ever getting hot enough to burn the fuel completely. The result is smoke, creosote, and low efficiency, regardless of many manufacturers' claims to the contrary.

Wood and CO_2

Burning any fuel releases CO_2. The idea that burning wood can be environmentally acceptable in terms of CO_2 emissions is based on the concept of carbon neutrality. This means that when wood is burned the CO_2 released is what the tree absorbed when it was alive, and what will be reabsorbed by the next tree that grows. This essentially means that no new CO_2 is being released into the atmosphere. When coal or oil is burned, the CO_2 that was "sequestered" in the earth millions of years ago is now released, adding to the total greenhouse gasses in the atmosphere. It is this release of old CO_2 that is seen as the problem, not the recycling of CO_2 currently sequestered in trees and plants. There is currently intensive study of the CO_2 balance, including the effects of cutting or harvesting too many trees or crops, and growing more species that sequester greater amounts of carbon.

Methane

We are not running out of natural gas. We just need to extract it from other sources. Methane, known chemically as CH_4, is the main component of natural gas. It is still being produced by the same type of bacteria that created the pockets of natural gas thousands of years ago that we are drilling into today. Newly produced methane can be mined from landfills or extracted from organic waste in an anaerobic digester. Because methane is a powerful greenhouse gas, it is doubly important to capture it from these sources.

Anaerobic Digesters

Anaerobic digestion is the process of extracting combustible methane gas from manure, sewage, and food processing waste. This promises not only to provide fuel for heating and generating electricity but also to assist in controlling the environmental impact of these waste streams. Extracting energy from biomass waste is part of an overall process to recycle these materials.

The anaerobic digestion process harnesses bacteria to do the work. Bacteria are responsible for the decay of dead vegetable matter, making its chemical components available to nourish new life. The first of two primary types is aerobic bacteria, which require oxygen to break down organic matter, and produce heat as a byproduct. This is the process at work in a compost pile. The second type is anaerobic bacteria, which break down organic matter in the absence of oxygen and produce methane as a byproduct. The anaerobic process occurs naturally at the bottom of swamps and boggy areas, and in the intestines of termites and large animals. It can also occur in your compost pile if it gets too wet and you don't turn it.

An anaerobic digester creates the same oxygen-free conditions found in a swamp, but the methane is trapped so it can be collected and used. Anaerobic digesters are most commonly found at sewage treatment plants for digesting raw sewage, on dairy and livestock farms for digesting manure, and at food processing facilities to handle process waste like cheese whey.

A digester uses either a large, above-ground tank or a covered lagoon for the digestion process. Depending on the climate where the digester is located, the tank may or may not need to be heated to maintain the proper temperatures for keeping the digesting bacteria alive and working. Where this is necessary, methane harvested from the digester is often used for the heating process. The manure, sewage, or other organic

Fig. 21. Anaerobic Digester at a Dairy Farm in Wisconsin. This system uses an above-ground tank for the digestion process. Photo by the author

waste moves through the digester over a period of about three weeks. During this time a complex bacterial process takes place to break down the waste, and the methane gas is collected. What is left, the effluent, is separated into liquid and solid components. Both are full of nutrients and can be used as fertilizer. Anaerobic digesters are primarily used to process waste into useable nutrients and to control odor and pests. The methane harvested is sometimes simply burned (flared), but using the methane from larger systems for generating electricity is becoming more cost effective as these technologies are developed.

The methane collected in the digester can be burned like natural gas in a generator to produce electric power. Because it does contain additional chemicals and impurities, digester methane can be corrosive, so special generators are used for electricity. It is also possible to clean up the methane so it can be added directly into the natural gas pipeline. Finding a cost-effective way of doing this would mean that methane from digesters too small to generate electricity economically could sell small amounts of gas directly. In the United States at present, only large-scale anaerobic digestion systems are seen as being economical for generating electricity. A farm digester large enough to produce the methane needed to generate electricity economically would need the manure from at least 500 cows. Cooperatively owned systems would work only if the farms were close enough together to avoid the high transportation costs for the manure.

With most successful farm digester projects, many of the benefits are not related to energy. The digestion process controls odor, pathogens, and weeds. Farmers can spread the digester effluent on fields anytime without causing unpleasant aromas. The dry component of the effluent is clean and almost odorless and can be used as bedding for the cows, saving the cost of other bedding material, or it can be sold as a garden soil amendment. The digester becomes part of the overall farm system, contributing to the business in a number of ways.

The federal government has recently instituted standards for waste management on large livestock farming operations known as Concentrated Animal Feeding Operations (CAFOs). These are also known as factory farms where thousands of animals and poultry are produced, which in turn produce mountains of manure. Regulation will attempt to control pollution of the air and ground water, rivers and lakes in the surrounding environment. Anaerobic digestion is seen as a potential solution to this waste management problem (in addition to producing useable methane for electricity generation), and a variety of government subsidies (both federal and state) encourage their use. However, the Sierra Club has taken a strong stand opposing public subsidies for CAFO digesters.[2] They feel these operations present a variety of threats, including pollution of air and water, animal cruelty, excessive use of antibiotics, and negative impact on local rural economies, and that public money equates with public support for these practices. They feel the technology has value for smaller operations but would like to see greater oversight of effluent quality and tighter regulation of system design.

Landfill Methane

In urban settings, anaerobic digesters are in common use at municipal sewage treatment plants as well as food processing plants and other facilities that produce organic waste. Many use the methane produced on site for controlling the temperature of the digestion process, but some generate electricity with it as well.

Landfills emit about 25 percent of the methane created by human-related activity in the United States. This methane is produced by the decomposition of solid waste and is now considered a notable energy resource. Capturing landfill gas not only makes it available for use, but prevents it from dispersing as a greenhouse gas in the atmosphere. According to the U.S. Environmental Protection Agency, Landfill Methane Outreach Program (LMOP), as of 2007 there were 424 landfill gas collection projects in 42 states, and another 600 landfills with collection potential.[3] Many of the successful projects use the methane for heat, while others generate electricity.

Transportation Fuels From Biomass

We are feeling understandably desperate to find eco-friendly substitutes for natural gas and gasoline, as these two fuels in particular do the heavy lifting in our everyday lives to provide heat and transportation. Biomass represents a straight substitution in some people's minds, and bioindustry hype encourages that point of view. This is particularly true for transportation fuels, where ethanol and biodiesel seem to promise us release from dependence on foreign oil. Naturally, it will not be that simple. First of all, our driving habits have been built on a glut of cheap oil that has been pumped from the ground. Biofuels will have to be grown and processed, and we will have to make decisions about how our finite areas of crop land will be used and how our forests will be managed. Our biomass resources will confront us with their worth. Do we eat them, or do we use them to drive to the health club?

The idea that our present transportation system, with its reliance on the private automobile, truck transport of goods, and airline travel, could remain exactly as it is except for replacing petroleum with biofuels is an unrealistic fantasy. Like our other sustainability challenges, there is no magic bullet for reducing and ultimately eliminating the use of petroleum for transportation. For one thing, we must simply use a lot less fuel. It is reasonable to assume, as James Howard Kunsler suggests in *The Long Emergency*, that we will travel less, eat more locally grown food, and live in more densely built communities.[4] Transportation is a huge subject of which energy plays only a part. For example, even if we find something to replace gasoline and keep driving as much as we do, we are still heading toward national gridlock.

Assuming that biofuels are only part of the solution, we can consider them to be an important part. At the present time there are two commercially viable biofuels: ethanol and biodiesel. These two fuels are currently being produced in many countries in the world. They are seen to have significant potential to expand the economies in developing tropical countries where biofuel crops are easily and cheaply grown. Sugar cane is one such crop. Brazil and the United States produce similar volumes of ethanol, with Brazil using sugar cane and the United States producing it from corn. Germany is the leading producer of biodiesel, producing it from rapeseed and sunflower oil. Biodiesel is also produced from waste grease and oil in Germany and the United States, making this the cheapest current biodiesel source.

Sugar cane produces the greatest volume of ethanol at 6,000 liters per hectare, followed by sugar beets at 5,000 liters, and then corn at 3,000 liters. The most productive biodiesel crop is palm oil, another tropical product, at 4,500 liters per hectare, followed by Jatropha (a hedge plant used for biodiesel in India) at 2,000 liters and rapeseed at about 1,000 liters.[5] Production of ethanol and biodiesel from these crops is commercially well established.

Concern in part that these fuel crops may ultimately compete with food crops has driven development of a new form called cellulosic ethanol. As the name implies, cellulose is the feedstock for this process, and it can be any number of things from wood waste to the organic portion of municipal solid waste to crop residue from farm fields. The key here is that cellulosic ethanol would make fuel from organic waste streams, relieving pressure on crop land. This form of ethanol requires the organic matter to undergo a liquefying process before it can be transformed into fuel, and such a process is not yet up to speed commercially.

There is considerable speculation on the subject of transportation fuels for the future. We are really anxious to be allowed to keep driving. Worldwatch Institute forecasts that biofuels could provide 37 percent of U.S. gasoline requirements within the next twenty-five years, and 75 percent if vehicle efficiency manages to double.[6] Some experts feel highly efficient electric cars are a better bet than a conversion to biofuels. Reducing our use of the internal combustion engine may not be a bad idea. There is recent research suggesting that, while ethanol use will reduce some pollutants, it will increase others, and burning ethanol may exacerbate the ozone problem in high-density areas.

Transportation is a highly complex issue in our very mobile society. Once again, the technology that will create alternative fuels cannot create solutions to other difficult transportation dilemmas. A variety of equity issues will still remain, such as assuring convenient access to good jobs and housing. Indeed, current economic disparities may increase as the cost of gasoline makes driving a car a real luxury once again. Also, the encroachment of urban sprawl on prime agricultural and forest land remains a central issue of sustainability planning. Urban traffic congestion is a major challenge to the revitalization of city centers. Public transportation and other options are still of vital importance regardless of the progress made to develop alternative fuels.

Other Non-Fossil Energy Sources

Hydrogen

The way some people talk about the "hydrogen economy" sounds a lot like what was said about nuclear power back in the fifties being electricity too cheap to meter. Hydrogen as the magic bullet is a great fantasy. Even though it may be possible eventually, the technology is not ready yet, and it might be a while before it is.

Hydrogen is the most common element in the atmosphere, but it is not really a source of energy in the sense that it can be collected, stored, and used. Because it doesn't occur in its pure form in any great quantities, it must be extracted from something else. To do this takes some other form of energy in just about equal amounts, so it turns out to be a trade-off. Currently the method used to produce hydrogen commercially, which is primarily used for the production of ammonia for fertilizer, is called steam methane reforming and uses steam to turn natural gas into carbon monoxide and hydrogen. Another method is electrolysis, which separates water into its component elements of oxygen and hydrogen by running a current through it. Obviously, using fossil fuel to generate the electricity for producing hydrogen by electrolysis does not eliminate the production of CO_2. Electricity from nuclear or renewable sources could address this problem, but these solutions introduce further complexities. There are also potential thermochemical processes for producing hydrogen that are in the experimental stages.

The bottom line, however, is that because of some inefficiencies in processing, the amount of energy in the extracted hydrogen is slightly less than the amount of energy used to produce it. Hydrogen doesn't offer significant advantages except where it can serve to store energy that might otherwise be wasted. For example, electricity produced by solar or wind power in an isolated area could be used to produce hydrogen that could then be piped to where it could be used. This is actually a way that solar power can be converted to power vehicles.

A number of technical difficulties remain. The storage of hydrogen is a problem because it does evaporate, and its density means that larger fuel tanks in vehicles would be required. The use of hydrogen in stationary fuel cells for generating electricity is likely to be in common use sooner than hydrogen vehicles.

Geothermal Energy

The word "geothermal" is in common use to describe two totally different technologies. The word itself means "earth heat." The technically correct use of the word describes the energy source from the earth's core, measured at 9,000 degrees Fahrenheit. In certain areas of the world, this heat comes close to the earth's surface and can be harnessed for both heating and generating electricity. For example, the geothermal resources in Iceland are famous for producing most of the space heating in that country.

Much of the geothermal energy resource in the United States is in the western part of the country. Geothermal energy currently produces about 8 percent of the electricity generated from renewable energy in the United States and is third behind hydropower (77 percent) and solid waste (9 percent). According to the Geothermal Energy Association, the United States leads the world in production of geothermal electricity, and with growing pressure to develop renewable energy sources more sites are being developed.

When a geothermal well is drilled, the hot water or steam that has been heated underground can make its way to the surface where it is used to drive generating turbines for electricity or employed in nonelectric processes as a heating source. The cooled water is then reinjected into the ground where it is reheated and used again. The first commercial geothermal generation plant in the United States was built in California in 1960 and remains in operation.

. Geothermal generation plants can operate twenty-four hours a day and produce no pollutants. Additional wells can be drilled on a modular basis as demand increases. The greatest disadvantage of this renewable source is its geographically limited occurrence. But even so, geothermal resources are currently harnessed in twenty-six states, and where these resources naturally occur considerable opportunity for development remains.

Nonelectric uses of geothermal energy range from heating greenhouses and aquaculture ponds to pasteurizing milk and washing wool. Space heating of buildings is another important application. There are close to three hundred communities in the western United States that use geothermal heat, many as district heating systems. Also, we can't forget the whole spa industry that promotes hot springs for health and relaxation.

Geothermal (Ground Source) Heat Pumps

The second use of the term "geothermal" refers to ground source heat pumps, which don't employ the heat from the earth's core but rather rely upon the even temperatures from 45 to 58 degrees Fahrenheit that are naturally maintained a few feet below the earth's surface. Some clean energy advocates claim that ground source heat pumps are not even renewable energy technologies because they don't actually draw on a source of renewable energy. Instead they simply use the earth as a heat storage device, or "heat sink." Regardless of what they are called, these systems offer a very efficient way of using electricity for space heating and cooling, and for water heating as well. They have tremendous potential for replacing natural gas for space heating in cold climates.

A heat pump uses a refrigeration cycle, like an air conditioner or a refrigerator. The refrigeration cycle is a process for transferring heat from one place to another using a chemical refrigerant that moves through a closed loop configuration. In a refrigerator or air conditioner, the heat is removed from the inside and transferred to the outside. The volatile chemical properties of the refrigerant make it capable of changing from a gas to a liquid form under pressure, within a narrow range of temperature. Within the closed loop, the refrigerant continually changes its state from liquid to gas to liquid again, absorbing and shedding heat in the process. The refrigerant turns into its gaseous form in the evaporator coils, where it absorbs heat. These would be the coils inside the walls of a refrigerator that remove the heat from the interior space. The refrigerant moves as a gas to the compressor where it is placed under high pressure. The compressor is what uses the electricity when the refrigerator is operating. From the compressor, the refrigerant condenses to a liquid in the condenser coils located on the exterior of the refrigerator, where it sheds its excess heat to the room. Finally, the refrigerant passes through the expansion valve where it once again assumes its gaseous state within the evaporator and absorbs more heat from the inside of the refrigerator. The refrigerator operates on a thermostat that tells the compressor when it is time to remove more heat from the refrigerator or freezer compartment, and the cycle is once again activated. A ground source or geothermal heat pump works in a similar manner except that it operates both ways, transferring heat from indoors to underground in the summer and performing the reverse action in the winter.

A central air conditioning system operates the same way, but the evaporator coil, located in the indoor unit, pulls heat from the air inside the home as it is circulated through the duct work to the unit. The condenser coil is located outdoors in a separate unit, and the heat from the house is dissipated by the condenser unit fan, which also uses electricity and is quite audible to the neighbors.

Heat pump efficiency is dependent on the difference in temperature between the indoor space and the underground piping loop, usually located nearby. There are also air-to-air heat pumps where efficiency is dependent on the temperature difference between the indoor space and the outdoor unit. These are most efficient in climates where the winter temperature does not drop below about 30 degrees Fahrenheit. Below that, there is not enough heat in the outdoor air to transfer indoors for an acceptable comfort level, and another source of heat such as a natural gas burner or electric heating must be added. However, ground source heat pumps use the underground temperature, which remains steady year-round. This makes them efficient even in the colder northern climates.

Ground source heat pump systems do not use an outdoor unit but rather contain all the refrigeration components in the indoor unit, usually the size of a standard furnace. The outdoor component of the system is a continuous loop of tubing that is usually buried vertically or horizontally in the ground. There is nontoxic glycol fluid circulating in this tubing that serves to absorb or dissipate heat as it passes through the indoor unit coil. As with a standard air conditioner, the ducts for a heat pump system circulate air through the coil and then back into the home or building.

These systems use very little electricity compared to standard electric heating and cooling systems because they rely on the consistency of the underground temperature and are not called upon to deal with extreme heat or cold. There are also water-source heat pumps that operate the same way, with the outdoor coils located in a pond, lake, or other body of water with sufficient depth to maintain consistent temperatures year-round.

Ground source heat pumps offer a great deal of flexibility in their installation. For example, they can be used for heating and cooling a school building using one large underground coil array connected to many small heat pump units for separate classrooms or other small areas as needed. These systems could also be installed for district heating and cooling, once again employing a large underground coil with separate units for individual buildings, adding on as required. The primary reason that ground source heat pumps are still relatively uncommon is that they have been competing with inexpensive natural gas. Installation of a ground source heat pump is often expensive, but demand has been rising with the cost of natural gas, and more heating contractors are gaining the skills needed to size and install these systems, which will drive costs down. As we develop more ways to produce clean electricity, ground source heat pumps will be efficiently heating and cooling more homes and commercial buildings without burning anything, something that seems almost impossible in cold, northern regions.

Hydropower

We have harnessed the power of moving water for thousands of years. It has been used to grind grain, saw wood, and run machinery like the spindles, looms, and sewing machines in the clothing mills along the rivers of New England. Hydropower has been used to generate electricity in the United States since George Westinghouse built the power station at Niagara Falls in 1896. Many small hydropower stations were built in towns across the country, and the federal government built the large dams like Bonneville, Hoover, and Glen Canyon. Hydropower provides 42 percent, or the second largest portion of renewable electricity used in the United States, behind biomass at 48 percent. The primary advantage of hydropower, beyond its lack of carbon emissions, is that its production can easily be controlled to keep pace with demand.

There are three basic types of hydropower generation. The first is called "impoundment," which is simply building a dam on a river to store water behind it. The water can then be released in a controlled manner to be channeled to drive the turbine that generates electricity. Dams are built for a lot of different reasons such as flood control, irrigation, or to create recreational areas. Very few of them generate electricity today. The second hydropower method is called "diversion," which is channeling part of the moving stream around a natural waterfall and harnessing only this portion of the river's energy. This is also called "run of the river" generation.

The third method, "pumped storage," is used to take advantage of low electric rates during late night hours. During the night, water is pumped from a lake or river up to a reservoir located above the water level. During the day the water is released to generate electricity that is worth more as peaking power when rates are higher. This method doesn't really qualify as a renewable energy technology, but it is a smart way of manipulating the costs of peaking power.

Hydropower is not a growing source of renewable electricity for a couple of reasons. First, there are only so many river locations where a hydropower plant can be constructed economically, and most of those have been used. Second, concerns have grown over the years that the damming of rivers for electricity generation causes damage to river ecosystems, disturbing spawning patterns and creating unnatural temperature patterns in the water. Some conservationists recommend removing dams where the topography has not been too dramatically altered. Many municipal hydro plants have been closed already simply because the local demand for power exceeded the plant's capacity and it was easier to build a new natural gas or coal plant.

Small hydropower systems still have a place in the clean energy picture, however, particularly for isolated locations or in combination with solar or wind energy. New "microhydro" technology has made it possible to capture energy from a river without affecting its natural attributes. A small amount of water can be diverted at the top of a waterfall and run through a microturbine to generate electricity, while the water is then released at the bottom of the falls to be on its way. These systems usually generate only a few kilowatts, primarily for use on site.

Tidal Power

In recent years there have been fascinating experiments with turbines that can generate electric power from tidal currents, making use of the movement of water between high and low tides. Because tidal patterns are influenced by the moon, we could call this "lunar power." Location is extremely important with tidal power installations. The best spots are those where natural obstructions cause swift currents, such as the entrance to a bay or river. Turbines are anchored to the bed of the river on concrete pilings in at least twenty feet of water, which must have a current of three knots or greater. Prototype systems have been built in the Strait of Messina in Italy and on the Gold Coast in Queensland, Australia. In the United States, Verdant Power has installed what resembles an underwater wind farm in the East River in New York City between Roosevelt Island and Queens, and the power they produce is sold to businesses on Roosevelt Island. Six turbines, which resemble three-bladed wind turbines, produce only 200 kilowatts of power, but the company has plans eventually to build up to 10 megawatts of tidal turbines. Tidal power is very dependable, and these underwater turbines also have the advantage of being out of sight, avoiding one of the most controversial aspects of wind power.

NOTES

1. W. P. Erickson, G. D. Johnson, M. D. Strickland, D. P. Young, Jr., K. J. Sernka, R. E. Good, "Avian Collisions with Wind Turbines: A Summary of Existing Studies and Comparisons to Other Sources of Avian Collision Mortality in the United States," *National Wind Coordinating Committee (NWCC) Resource Document*, August 2001.

2. Sierra Club, *Sierra Club Guidance: Methane Digesters and Concentrated Animal Feeding Operation (CAFO) Waste*, October 20, 2004, http://motherlode.sierraclub.org/ MethaneDigestersSIERRACLUBGUIDANCE.htm, downloaded October 4, 2007.

3. U.S. EPA, *An Overview of Landfill Gas Energy in the United States*, http://www .epa.gov/lmop/docs/overview.pdf, downloaded October 4, 2007.

4. James Howard Kunstler, *The Long Emergency: Surviving the End of Oil, Climate Change, and Other Converging Catastrophes of the Twenty-first Century* (New York: Grove Press, 2005).

5. Worldwatch Institute, *Biofuels for Transportation: Global Potential and Implications for Sustainable Agriculture and Energy in the 21st Century*, 2006.

6. Worldwatch Institute, *Biofuels for Transportation: Global Potential and Implications for Sustainable Agriculture and Energy in the 21st Century*, 2006.

RESOURCES

U.S. Government Websites

U.S. Department of Energy, Energy Efficiency and Renewable Energy (EERE)
This website offers information about a broad range of technologies and programs for individuals, businesses, and institutions.
http://www.eere.energy.gov

National Renewable Energy Laboratory (NREL)
Located in Golden, Colorado, this is the U.S. DOE laboratory dedicated to renewable energy technologies. This site offers information about the latest technologies and government research.
http://www.nrel.gov

NREL wind resource maps
http://rredc.nrel.gov/wind/pubs/atlas/maps.html

Sandia National Laboratories
This U.S. DOE national laboratory focuses on national security technologies, but they have a well-known renewable energy and distributed generation laboratory. This website offers a different view of renewable energy.
http://www.sandia.gov/Renewable_Energy/renewable.htm

Nonprofit Websites

Union of Concerned Scientists
An independent nonprofit, this organization is known as a reliable source of independent scientific analysis of environmental and energy issues. It contains some excellent information about renewable energy technologies.
http://www.ucsusa.org

Websites For Specific Renewable Energy Resources:

Solar
American Solar Energy Society (ASES) and International Solar Energy Society (ISES)
ASES was founded in 1954 as a nonprofit organization to promote solar energy, energy efficiency, and other sustainable technologies. ASES is also the American chapter of the International Solar Energy Society (ISES).
http://www.ases.org
http://www.ises.org/ises.nsf!Open

Solar Energy Industries Association (SEIA)
This is the nonprofit trade association for solar energy professionals. Their website offers information about the business of solar energy.
http://www.seia.org

Wind
American Wind Energy Association
A nonprofit information and advocacy organization promoting wind power as environmentally sound, economical, and secure.
http://www.awea.org

Windustry
A Minneapolis-based nonprofit promoting wind energy to rural communities. Their website includes information about community wind projects.
http://www.windustry.org

Biomass
Woodheat.org
This non-profit Canadian website promotes the use of wood for heating homes.
http://www.woodheat.org

U.S. Environmental Protection Agency, Clean Burning Wood Stoves and Fireplaces
A compendium of information about wood as a heating fuel, including links to other organizations.
http://epa.gov/air/woodstoves/partner.html

Geothermal Generation
Geothermal Education Office (GEO)
An educational nonprofit organization founded in 1989 offering information about geothermal electric generation.
http://geothermal.marin.org

Ground Source (Geothermal) Heat Pumps
Geothermal Heat Pump Consortium

This is an educational nonprofit promoting the use of ground source heat pumps. Their website offers good basic information and informative graphics.
http://www.geoexchange.org/

Tidal Power
Verdant Power
This is the website of the energy company building a tidal turbine farm in New York's East River. Their website includes an animation of a tidal power turbine farm.
http://www.verdantpower.com/

CHAPTER FIVE:

Community Energy and Sustainability

The overview presented in previous chapters focused on our relationship with energy resources. But these resources, and our use of them, are not separate from community issues such as loss of farmland to encroaching suburbs, air and water pollution, the affordability of housing, or employment opportunities in the local economy. These and other areas of concern that relate to maintaining our quality of life are usually addressed in discussions of "sustainability" or sustainable community planning. Sustainability planning integrates solutions to environmental, social, and economic challenges in mutually beneficial ways with the goal of preserving or improving our quality of life.

It is important to note that many of the "unsustainable" aspects of our lives relate directly to our use of energy and our acquired dependence on fossil fuels. Reducing emissions from power plants and vehicles will mean cleaner air and water. Building wind farms and using biomass transportation fuels can create new economic opportunities for farmers and rural communities. Green building standards are creating a whole new industry of energy efficiency technicians, installers, and products. Understanding the connection between energy and sustainability helps us integrate our approach and gives us a whole new set of tools for finding solutions.

Dramatically reducing our use of fossil fuel will relieve much of the current environmental stress that is caused by pollution and potential global warming. About 40 percent of the carbon dioxide emissions generated from fossil fuel combustion comes from the generation of electricity, followed by about a third from all sectors of transportation. The remaining 27 percent comes from other industrial, commercial, and residential uses of fossil fuels, mostly for heating. Arriving at energy sustainability will mean finding strategies to use less energy as well as finding efficient, renewable ways to produce more.

CO_2 reduction is more than integral to all we think of as sustainability, it is paramount. Climate change experts are now recommending that humans reduce CO_2 emissions by 80 percent over the next thirty-five years in order to prevent truly grave damage to the earth's atmosphere. Even nations already more involved in CO_2 reduction strategies than the United States find this a daunting goal. Although we've begun to recognize the need to find a sustainable energy path for the future, per capita use of energy in the United States continues to rise. Yet, while reducing fossil energy use in our homes and vehicles is vitally important, a glance at the overall picture shows that we as separate individuals can only do so much. National and state-level policies, laws, and programs are also a necessary part of the solution, but for most of us it is difficult to relate to these broad strategies, and we don't generally feel we can be part of making them happen. Reducing the carbon footprint may be a global goal, but it can only be achieved at the local level. We need to work together where we can be most effective and where we are already addressing many related sustainability issues—in our communities.

Integrating Energy into Sustainability Planning

Local governments are ideally positioned to coordinate energy concerns with such issues as land use, transportation, housing, and environmental quality, even though this role is a new one for municipalities and potentially challenging to implement. For communities, an important step will be making efficient use of energy resources a common practice in all design and planning processes. If a community has already begun giving sustainability principles political priority, integrating clean and efficient energy use can only strengthen the effort.

Cities and towns in every state are already addressing sustainability concepts, many in response to the U.S. Mayors Climate Protection Agreement. If fossil fuel energy use is responsible for most of the CO_2 emissions we can control, then any discussion of climate protection must include a significant energy component. There are plenty of approaches to take, although some emphasize energy solutions more than others. Municipalities can choose "smart growth," "green development," or "sustainable community," or they can decide to become a "solar city" or pledge to reduce their "carbon footprint." There are also established design approaches like New Urbanism, or the USGBC's latest rating program, LEED for Neighborhood Design or LEED-ND®. The Natural

Step, a program highly successful in Sweden and beginning to attract notice in the United States, not only features community energy strategies, it is also very a effective process for inspiring citizen involvement.

Energy Self-Reliance

Many communities are becoming interested in the concept of "energy self-reliance." This idea is appealing because it implies a certain benchmark of stability for the local economy, the natural environment, and the safety and security of the community. The concept of energy self-reliance might sound like simply taking the whole town off the electric grid and existing in energy isolation from neighboring communities. To do this would require building enough electric generation to guarantee that all needs would be met, even during emergency peaks of demand like overly hot summer afternoons when everyone's air conditioner is running. For example, with a dependable supply of wood waste from forestry or manufacturing, a wood-fired generation plant could economically produce the base load power needed for a small town. This plant would provide power through the local distribution system, avoiding the costs of a transmission interconnection and gaining security advantages.

However, it is likely to be far more economical to remain connected to the grid while maintaining a state of net zero energy. This means that local electricity needs are met by local generation, with the transmission grid used as a backup storage battery. When the local systems are producing more power than needed, it goes into the grid, and when they are not producing enough, power is drawn from the grid. Net zero energy means that over time, usually a period of a year, the plus and the minus even out to zero. The state of net zero energy would then equate with energy self-reliance.

If a locality has fossil fuel resources to tap and its only priority is to avoid importing generation fuels from elsewhere, it could meet its goals for energy self-reliance by using them. However, if community goals include reducing greenhouse gas emissions as well, local coal, oil, or natural gas would not be eligible. Renewable resources available locally could be the sunshine that falls on building rooftops, the methane produced by the local landfill, livestock farm, or sewage treatment facility, the wind on local hilltops, or crop waste and forest biomass that could be burned in a generation plant. All sizes of renewable energy systems contribute to energy self-reliance, from residential solar panels to a biomass combustion plant, a commercial scale wind farm on the edge of town, or hydrogen fuel cells powering the post office.

Energy self-reliance for transportation is much tougher to achieve, but land use decisions that give denser development and public transportation priority reduce the need for petroleum imported from outside the local area. Biofuels and electric vehicles could further reduce local needs. However, the fact that we import most of our food, clothing, and other essentials from elsewhere illustrates how the issue of local energy self-reliance must also be addressed at the regional and national levels.

Making a Plan

We already have the technologies we need for renewable generation to create net zero energy communities, and whatever we don't have is probably being developed or dreamed about. The new technology side of the equation is the easy part. It is more difficult, as always, to reduce demand in order to make the supply side more economical, particularly as our towns and cities grow. A plan to achieve net zero energy must be very aggressive about energy efficiency and conservation in all sectors. This is the part that requires community participation and commitment, from homeowners, business owners, industry, private institutions, and the municipal government itself. There are also collaborative projects such as district heating and cooling systems, or cogeneration systems that capture and use waste heat, which can create efficient energy use at the community level.

Most efficiency or clean energy projects at the state and local levels address energy use in a particular sector, such as compact fluorescent lights for homeowners or HVAC maintenance training for commercial building managers. Traditionally, energy efficiency and renewable energy have been promoted by separate organizations or government agencies through different programs and with different goals. From the consumer perspective, we usually turn to our utility for energy efficiency programs, while we seek much of our information about renewable energy technologies from grassroots, nonprofit organizations or specialized government programs. Both efficiency and renewable energy advocates have further diffused their overall impact by targeting individuals, businesses, or institutions separately, rather than addressing whole communities. Furthermore, energy professionals from utilities, government energy offices, or nonprofit advocacy groups have traditionally worked in isolation from urban planners, architects, and community sustainability activists, who each have their own particular concerns.

Integrating energy into sustainability planning works best as a collaborative process among all these disciplines because it not only broadens the information base, it introduces the potential for new professional relationships. Energy planning can be inserted into an existing process such as comprehensive planning, or carried out independently to complement other planning goals. Community level planning coordinates programs and projects designed for individual sectors, thereby identifying more opportunities both to reduce energy use and to raise awareness than are available through scattershot efforts. Some energy goals are most appropriately addressed at the community level such as planning for energy infrastructure and security. The best kind of plan is one based on the resources and needs specific to the community, one that is created by a participatory process that includes the whole community, and one that is planned as a living document, with built-in flexibility to allow ongoing changes. It is also a plan that is designed to include implementation strategies and methods for measuring progress.

Net Zero Energy Through Smart Growth

"Smart growth" planning typically looks at land use through the lens of sustainability by integrating a variety of community concerns, from environmental quality to economic development and social justice. Smart growth planning goals might include tightening the urban fabric through redevelopment of land with existing infrastructure, promoting transportation options, providing affordable housing, and making employment opportunities accessible. Smart growth planning also seeks to protect farmland and forests, wildlife habitats, open spaces, and ground water resources, along with preservation of cultural and historic sites and revitalization of community identity. These are all goals generally associated with sustainable community planning. Smart growth planning does not generally incorporate specific strategies regarding energy production or use. However, we can identify areas where energy issues intersect with smart growth land use goals and stretch the scope of the typical comprehensive smart growth plan beyond land use into more of an overall vision for the future of a community.

Energy and Smart Growth Planning

The balance of this chapter uses mandatory and optional elements from the model statutes in the American Planning Association's *Growing Smart^{SM} Legislative Guidebook* as a template to illustrate the role energy plays in smart growth planning.[1] The organization drafted its *Legislative Guidebook* in 2002. This guidebook was developed to assist states in drafting smart growth planning legislation that would be sufficiently detailed to create effective plans, allow local plans to fit within regional ones, provide ways to integrate plan elements within plans, and establish citizen participation and ongoing plan evaluation processes. The model statutes offered in the guide are presented as mandatory and optional comprehensive plan elements. We will explore these elements: Issues and Opportunities, Land Use, Transportation, Community Facilities, Housing, Economic Development, Natural Hazards, Critical and Sensitive Areas and Agriculture, Forest, and Scenic Preservation, and Program of Implementation. Finally, to distinguish our version, we will call it an "energy-conscious" smart growth plan.

Issues and Opportunities

The Issues and Opportunities element typically lays the foundation for the rest of the comprehensive plan. This is the element that spells out the objectives, policies, and vision of the community along with challenges and potential problems. It also provides space for demographic background information and other statistical data that describe the community including employment opportunities, housing stock, educational and recreational resources, and population trends and forecasts for age and income level demographics that can be used in the comprehensive planning process over a twenty-year period. There is nothing in this element that refers specifically to sustainability, but it does provide the opportunity for a community to present a profile of itself and its goals for the record. If a town wishes to make sustainability or environmental quality part of its future vision, this is the place to mention these goals, including any applicable timetables.

A municipality could easily include a vision or goals related to sustainability that it might already have established such as signing the U.S. Mayors Climate Protection Agreement or pledging to build a green industrial park. General plans for achieving such goals could include reducing fossil fuel use through energy efficiency or becoming more energy self-sufficient on a local level by encouraging renewable energy use. Other potential community goals might be reducing energy use and installing renewable energy technologies in municipal facilities, developing new businesses based on locally grown energy crops like corn or switchgrass, or developing a wind farm project through the municipal utility.

Comprehensive plans often contain some history of the town along with other information that outlines its character and aspirations. There is no reason why the snapshot of town demographics in this element couldn't be expanded to include a subheading describing energy assets as well, like available wind, solar, or hydropower resources. Perhaps there is an abundance of certain crop waste, manure, or forest resources, or maybe there is a source of waste heat from a generation plant or manufacturing facility that could be harnessed to become a community asset. There might also be members of the workforce whose skills could be identified as useful in a biodiesel plant or a wind turbine manufacturing facility, such as displaced workers from industrial plants that have closed.

Land Use

This is the key element for any smart growth plan because its purpose is to interpret the community vision in physical form. The Land Use element looks specifically at how land is currently being used and then establishes guidance for development and redevelopment of both public and private property in the future. Presumably, this is where a municipality can put its communal foot down about sprawl. The *Legislative Guidebook* recommends including descriptions of current land uses, building densities, and property value, but also suggests projecting population and growth area distribution, areas for economic development, and the location of prime crop land and environmentally sensitive land areas such as wetlands and floodplains.

The Land Use element provides the opportunity to incorporate the net zero energy concept into the smart growth vision itself. The smart growth planning process will likely include revisiting previous approaches to land use to identify where regulations began encouraging sprawl. It is now accepted practice to offer more flexibility with regard to mixed use and high density developments, streetscapes, and community spaces. In other words, smart growth planning is as much about eliminating the roadblocks in zoning regulations and ordinances that encourage sprawl as it is about creating new rules. It is not unusual to find zoning practices that discourage energy efficiency and renewable energy. Two examples are minimum lot sizes that make more driving mandatory and setback requirements that forbid duplex construction on adjoining lots with a common wall between that can cut energy bills for both households. The smart growth flexibility that allows for higher densities usually improves both building and transportation energy efficiency as well.

To make their plans energy conscious, local governments also need to revisit any ordinances or rules related to residents or businesses installing renewable energy systems. There were many ugly systems installed in the 1970s, which caused both neighborhood associations and city authorities to make rules and pass laws against them. Times have changed and system appearance has improved. Solar thermal and solar electric panels are no longer regarded as exotic or radical by most people. Even so, negative prejudice against them remains in some minds, and conflicts still arise. If a community establishes a goal of net zero energy use for itself, then a commitment to promoting renewable energy systems is essential to reaching that goal. Laws still on the books that support a negative point of view should be revised or repealed. This is particularly true of permitting processes for renewable energy projects. Cumbersome application requirements and costly permitting fees are identified by renewable energy installers in many areas to be significant roadblocks to wider adoption of home-scale systems.

New rules and guidelines that promote and encourage renewable energy use are also important. For example, solar orientation and access are two potential concepts to include in renewable energy–friendly zoning and building codes, along with requirements that new homes be made solar ready. For local governments with jurisdiction over rural and agricultural land, supporting renewable energy technologies through land use will also include zoning and ordinances regarding wind energy, anaerobic digestion, and biomass fuels, as mentioned in the Agriculture, Forest, and Scenic Preservation element.

Projections for future land use should also take into consideration the energy needs of additional housing, commercial, or industrial development, particularly if the community is interested in achieving a sustainable net zero energy state. This element might require new developments to support the community's own energy needs through renewable energy generation on site or through financial investment in green power that contributes locally or regionally to the grid.

Transportation

We have become such a mobility-oriented society that as a planning issue, transportation has become a many-headed monster. The *Legislative Guidebook* acknowledges that a transportation element could address a number of things including traffic circulation, mass transit, and air, water, and rail transportation, as well as bicycling and walking alternatives, and, of course, parking. It suggests that comprehensive plans deliver guidance on how the local government will deal with the various modes of private and public transportation, and how it plans to coordinate with state and regional highways and other transportation corridors. Although it does mention compliance with the Federal Clean Air Act, particularly for localities within nonattainment areas, the *Legislative Guidebook* includes no language that directly addresses reducing the use of fossil transportation fuels to reach local greenhouse gas emission goals. The use of petroleum for transportation has a significant impact on the environment, producing about one-third of human-generated CO_2 emissions. Clearly, ways to reduce our use of it should be introduced into this element.

As with any other energy sector, the two ways to reduce the environmental impact of transportation fuel are either to use clean fuel or to use less fuel, and applying both strategies is the optimal path. If a community wants to encourage the use of biofuels such as ethanol or biodiesel, the energy-conscious plan could set a timetable for replacing the municipal fleet with flex-fuel vehicles. It could also establish incentives to attract a biofueling station to town for local customers, and join with regional and state officials to build the network of stations it will take to make these fuels available for the many flex-fuel vehicles already on the road. Natural gas is also considered an alternative fuel and natural gas vehicles are used in a number of cities, but future fuel costs are sufficiently unstable to warrant caution in adopting this technology. However, it may eventually be economical to process agricultural or landfill methane and inject it into the natural gas pipeline, potentially stabilizing the price.

Another strategy for reducing the use of gasoline is adding hybrid or fuel-efficient vehicles to the municipal fleet, particularly for staff that travel beyond the biofuel network. Energy efficiency also enters into other aspects of transportation. The concept of synchronizing traffic lights to reduce idling traffic has been used for many years to reduce both fuel use and exhaust emissions. Some cities use traffic sensors at intersections to change the stop lights more or less often when traffic is not heavy, or to employ flashing red signals or other traffic design strategies to prevent unnecessary idling. Even fuel-efficient vehicles get zero miles-per-gallon when stopped at a light.

Seeing the Light

Traffic lights and controls use electricity, and communities are taking steps to reduce the environmental and economic expense of these as well. Light Emitting Diodes (LEDs) have been used for a long time as control panel indicator lights in electronic equipment. They are highly efficient electronic lighting that produces little or no heat and lasts much longer than incandescent lights. Recently, lighting manufacturers have begun using LEDs for many standard lighting applications including traffic lights, and many cities are beginning to replace incandescent traffic lights with LEDs. Hundreds of LEDs are packaged together with reflectors that produce the light levels and intensity required for traffic lights. Directional arrows require fewer LEDs. Energy savings can be 80 to 90 percent, which will save both electricity and a chunk of the municipal budget.

The Smart Growth Energy Bonus

Smart growth approaches to transportation planning tend to save energy as a bonus. Higher density development reduces vehicle miles traveled (VMTs) by making it possible to walk or ride a bicycle instead of drive. A well-developed public transit system that goes where and when people need it not only increases mobility for those who don't drive but reduces pollution and energy use as drivers leave their cars at home. The development of bicycle and pedestrian pathways, and establishment of "park and ride" locations or special traffic lanes for carpooling commuters are other ways to increase the options available while saving energy.

Community Facilities

This element describes future plans and goals for all the physical facilities owned by the city or town and upon which it relies to continue functioning. These include everything to do with managing waste, from sewage treatment to storm water management, landfills, and recycling. Municipal facilities also keep the streets clean and the urban forest pruned. They provide for the public safety with police and fire departments, and for the public health through their health departments, hospitals, and various care facilities. There are schools, libraries, and community colleges that serve the community at large. Municipalities also support their neighborhoods with parks, community centers, and athletic facilities such as pools and ball fields. City hall oversees the physical character of the community's private sector as well through building permits, zoning, and the comprehensive planning process itself. The Community Facilities element of the comprehensive plan is central to the smart growth vision because it is here that all the planning and politics manifest themselves in physical reality, and the community's money and resources are invested for its future quality of life.

Like other government bodies, local governments have traditionally played two influential roles. One is to make the rules, through laws, fees, and taxation, and the other less obvious but vital role is that of being a consumer of goods and services itself. This latter role contributes economic impact to the community but it can also exert considerable influence by example. To implement a successful energy-conscious smart growth plan that rings true within the community, a municipal government must walk the talk. The Community Facilities element is where the municipality can be addressed as a consumer, not only with regard to the resource issues implicit in its land use planning, but more directly in what it buys, what it uses, and how it takes responsibility for the waste it produces.

Municipalities themselves use a great deal of energy to keep things running, from sewage treatment and waste management facilities to police and fire departments, schools, water utilities, telecommunications, and city hall itself. There are the obvious uses like lighting for buildings, traffic control and outdoor facilities, space heating and cooling, gasoline for police cars, and the diesel fuel used to run garbage trucks and fire engines. But there is also the electricity required to run all the water and sewage pumps and their computerized controls, police and fire communications systems, and specialized equipment in health facilities. Cities also need gasoline to mow all the municipal lawns, chip their dead trees, and fuel their animal control fleets.

An energy-conscious smart growth plan looks at making energy use as efficient as possible in all community facilities. The adoption of green building standards addresses new construction and renovation of existing municipal buildings. Purchasing guidelines for energy-efficient equipment like lighting and mechanical systems can encourage life-cycle costing, which frequently shows that their installation will pay for itself within a few years if not just a few months. Municipalities can utilize performance contracting to help finance efficiency upgrades because the equipment is paid for from projected energy savings.

All of these approaches need good preliminary data on how much the government is currently spending on energy. Therefore, an essential goal in this element is the gathering and coordinating of information about energy use among departments and facilities. Input from those who actually drive the trucks, maintain the equipment, and manage the buildings is invaluable. The need to gather this information presents an excellent opportunity to invite participation in the process. It can also reveal situations where departments or offices are working at cross-purposes in meeting energy-conscious smart growth goals.

Renewable Resources

Municipalities also have renewable energy resources available. The sun shines everywhere, and even in cloudier climates it can provide energy for heat or electricity. Wind energy resources vary from location to location but are often worth investigating for electric power. Most hydroelectric resources have been harnessed or they have been abandoned for economic or environmental reasons. However, new low-head and run-of-the-river hydro technologies might make the local waterway an economical energy producer once again. Geothermal resources for electricity generation or heat are available in some parts of the country, particularly in the West. Geothermal or ground source heat pumps work everywhere, and because they provide low-cost heat using electricity they are effective in colder climates as a substitute for natural gas. Their ability to provide low-cost cooling in warm climates is a boost for efficiency. A professional engineering assessment of these resources is worth the investment.

Waste as an Energy Resource

Communities also create different kinds of waste, which the municipal government has the responsibility to manage. The challenge is to consider as little of it as possible to be waste by finding other useful things to do with it. Many "reduce, reuse, recycle," and municipal yard waste composting operations programs are already working to use landfill space more efficiently. However, other forms of municipal waste can be considered sources of renewable energy, such as the methane that can be extracted from landfills and from anaerobic digesters at sewage treatment plants. These units are typically used to generate electricity, usually to provide power on site. In Europe, small methane digesters are located outside restaurants and food processing plants and the methane is used on site for cooking. It would not be surprising to see this technology adopted in the United States, particularly at institutional settings like school campuses or correctional facilities that produce a great number of meals every day. Depending on the size of these digesters, it might also be economical to use the methane to generate electricity.

Another source of municipal waste with energy potential is wood from construction, demolition, and urban forestry management. There is also wood waste from private industry that can become a significant burden on the municipal system. Depending on the local supply, wood waste can economically fire heating boilers for schools or be used to fuel a local generation plant. Urban forestry residue provides a clean and steady source of wood. In any city or town, a certain number of trees in parks and medians need to be pruned or removed every year. Some of those trees are chipped for garden mulch or compost, but not all. A major source of wood waste from private industry is used shipping pallets. In some manufacturing centers these have become a waste disposal problem even though there is a certain demand for them as the raw material for decorative garden mulches.

Waste wood from construction and demolition is more expensive to acquire, although new methods for managing these wastes on site are changing that. In the LEED rating system it is advantageous to keep as much construction site waste as possible from going to the landfill. New wood waste is usually separated and sold for processing as fuel or mulch. Demolition waste is trickier because of toxic paints and varnishes and the need to remove nails, screws, and other old hardware. Construction and demolition waste currently constitute a large portion of municipal solid waste. As communities discover the value in this waste wood, more economical ways of obtaining it will no doubt be developed.

There are two more sources of municipal waste to look at. One does not seem like an energy source at first glance, and the other would appear to be one, but is not sustainable in the long run. These sources are waste heat and municipal solid waste. Power plants are a major source of waste heat because a typical coal combustion plant is about 35 percent efficient. This means that only about a third of the energy in the coal is being converted to electricity, and the rest is lost as heat. Combined heat and power (CHP) plants take advantage of the waste heat by capturing it and using it for other purposes such as space heating, hot water, or a variety of industrial process that need heat. Most coal-fired generation plants do not harness the extra heat, and it goes up the stack to dissipate as waste heat. Some generation plants use a river or lake as a source of cooling, in which case the higher temperature of the discharge water can adversely affect the natural habitat in the water.

Generation plants are not the only sources of waste heat. Manufacturing facilities are also potential sources, particularly those that include a drying process in their production. It is often worth investigating the proximity of waste heat sources to areas where new development might be able to use the energy for space or water heating. The City of Holland, Michigan, tapped the waste heat from its local generation plant to heat its downtown sidewalks and keep them free from snow and ice, making pleasant tourist shopping a year-round experience. District heating and cooling systems can also use waste heat to economically serve high density urban neighborhoods or large hospital or academic campuses. The Midtown Eco Energy project in St. Paul, Minnesota, is a combined heat and power plant that will generate electricity and provide heat for customers within a 1.5-mile radius of the plant. It will use waste wood as fuel.

Municipal solid waste (MSW) is regarded as a renewable energy source by some state and local governments because burning it involves no new energy, and much of it is organic food and yard waste. However, this designation is questionable, because a portion of the municipal waste stream is made from, or with the use of, fossil fuels. According to the U.S. EPA,[2] there were eighty-nine waste-to-energy (WTE) plants in the country in 2006, producing about 2,500 megawatts of power. Waste-to-energy plants became popular in the eighties as a way of managing high volumes of municipal solid waste in the face of shrinking landfill space. Since then markets for recycled materials have grown. Many municipalities have banned yard waste from garbage pick-up and created separate composting programs. The U.S. EPA now recommends that municipalities employ these and other strategies for using a greater percentage of the municipal waste stream for higher value end uses before resorting to burning it. Also, because it contains unknown quantities of toxics such as chemicals and heavy metals, MSW must be burned in plants equipped with expensive emission controls.

It may seem to be a great idea to burn our waste to make electricity instead of just burying it. However, burning something for its energy is the least valuable thing we can do with it. The path toward keeping materials out of the landfill by reusing, recycling, and composting is a far more sustainable course. Furthermore, if we kept building expensive WTE plants, we would have to keep throwing away vast quantities of garbage to keep them running. This would not help us much to reduce the impact of our energy-intensive consumption patterns.

Utilities

The *Legislative Guidebook* includes mention of public utilities under the Community Facilities element, including the energy utilities that provide gas, electricity, or steam. Municipal utilities provide many services. Most commonly they include water, electricity, and sewage treatment, but many provide natural gas, steam, communications services, or cable TV. These are all considered services that contribute to the communal well-being.

When utilities are owned and operated by the municipal government, there are opportunities to incorporate their goals into the community's energy-conscious smart growth plan. Even for privately owned utilities, efficient land use planning that encourages infill and higher density development will also mean efficient energy use in providing utility services, in both their construction and operation. Internal efficiencies and energy innovations undertaken by either public or private utilities can augment the benefits of efficient land use planning. Energy utilities in particular need to be in on the planning conversation, because their contribution and participation are essential to meeting sustainability goals, particularly those related to reducing carbon emissions. However, other utility services can account for a significant portion of the municipal energy bill, and utility representatives should be at the table simply for that reason.

Among utilities, municipal electric and gas utilities and rural electric cooperatives already have a stake in their communities. A municipal utility is generally owned by the local government to serve its citizens, and the coops are owned by their member/customers. They share a democratic, customer-oriented approach that is quite different from the stockholder-driven, investor-owned utility model. Coops and municipals are locally based and usually small, which gives them greater flexibility to try new approaches, particularly when their customers and members want them to move in that direction. These publicly owned utilities are in the ideal position to offer leadership on new utility models that promote renewable energy generation and some serious demand side management as part of smart growth comprehensive planning goals.

Many investor-owned utilities are becoming more responsive to concerns about climate change and the carbon footprint. They see the regulatory writing on the wall in the inevitability of environmental controls on CO_2 emissions and potential carbon taxes or cap and trade programs. Also, building new generation plants is becoming more difficult and expensive. However, the largest investor-owned utilities frequently have little interest in getting involved at the community level with local energy planning. When the whole industry got leaner in the eighties and nineties, utilities closed many of their local offices, cutting off their direct customer connections. This is not to say it is impossible to work with one of these utility giants, but it can be difficult to establish a fruitful working relationship. It seems as though, the smaller the service territory, the more likely an investor-owned utility will be to get involved in a community planning exercise to tighten up energy use.

Every effort should be made to involve utilities in the community energy planning process. They have tremendous expertise in providing energy services, and they can also offer public education programs and other support for meeting sustainability goals. However, utilities are not accustomed to a shared planning process. Traditionally, utilities plan by projecting existing rates of growth in population and per capita energy use into the future and then laying plans for new generation plants based on these projections. It is in everyone's best interests to begin collaborative planning for future generation development, along with efficiency and renewable energy goals, and strategies for land use to promote energy efficiency. For the electric utility it can mean saving costs for capital investment and perhaps diversifying its generation portfolio, and the community will clearly understand it has the choice to cut down on energy use in order to prevent a new power plant appearing nearby.

Housing

As with other elements, the *Legislative Guidebook* does not directly address energy issues in its housing element. However, it does establish affordable housing as an important community concern, and energy use is directly related to how affordable a home will ultimately be. Energy-conscious smart growth issues related to housing include efficiency of the units themselves, density of units in neighborhoods and mixed use developments, and orientation and installation of renewable energy technologies for both electricity and heat.

An energy-conscious smart growth plan would require the gathering of energy data on residential buildings in the community, based on the age, type, and condition of typical dwellings, to determine an approximate level of energy efficiency in the existing housing stock. It would also be relatively easy to assess the solar orientation and tree cover in residential neighborhoods. This data would be useful for deciding which strategies would be most economical for the community to pursue, and which could be used to support programs that promote energy-related objectives such as those in the U.S. Mayors Climate Protection Agreement.

The energy efficiency of new homes or multifamily buildings is easier and cheaper to accomplish than that of existing units because it is built in from the start. Municipal governments frequently have the authority to adopt efficiency standards for home construction such as the International Energy Conservation Code or ASHRAE 90.2, if such standards are not already put into place at the state level. Voluntary programs like

ENERGY STAR have a proven track record with home builders and can help establish strong public/private collaborations in meeting community sustainability goals. The USGBC LEED standards for residential construction go beyond energy efficiency to reducing other environmental impacts on the site and on occupant health. The new LEED-ND rating was created for the design of whole neighborhoods that achieve similar sustainability goals. Comprehensive plan criteria for affordable housing could require not only that dwellings should be for sale at a reasonable price, they should be economical to maintain as well, and this means energy costs in particular. Pricier homes could also be required to perform well. They typically use more energy than more modest homes, and we can no longer afford to waste energy, no matter how much money we have.

Few argue with the concept of energy-efficient homes, but adding renewable energy can be controversial. There have been cases of homeowners from across the country having to fight with their neighborhood association, a local government official, or utility staff about installing solar panels or a wind turbine on their property. Clear commitment at the municipal government level to clean energy sources is essential if they are to make an impact on reducing carbon emissions. These are new and unfamiliar technologies to many people. Proactive local government support and accurate information made available to everyone can go a long way toward clearing up common misperceptions. This is an area where a community energy planning process can be essential in promoting the clean energy sources that will end up benefiting us all. For example, the Natural Step program begins with participatory community education about how our energy use affects the environment and why we need to reduce its impact. Proponents have found it is easier to gain cooperation for planning and meeting long-term energy sustainability goals when people understand both the benefits and what is at stake.

When energy-conscious smart growth planning adopts standards and objectives to encourage use of renewable energy, education efforts should include municipal staff as well, particularly permitting officials and building and trades inspectors who will be enforcing energy codes. An electrical inspector should understand everything involved in the interconnection of a solar electric system, and a plumbing inspector should be able to recognize what's required for a quality solar hot water system installation. Planning departments need to understand the principles of solar orientation for neighborhood developments.

Local governments could sponsor technical training on energy efficiency and renewable technologies for builders and developers that would enlist them in meeting community goals while attracting home buyers. This is where collaboration through ENERGY STAR would lend particular support. The city could sponsor training for real estate developers in such subjects as life cycle costing, solar-ready homes, and the economics of solar hot water systems for multifamily buildings. Municipalities can also work with local financial institutions to promote energy-efficient mortgages that take into account the lower monthly energy costs of an efficient home.

Economic Development

This element looks at the nature of the local economic structure and work force, along with potential employment and new business opportunities, employing the smart growth plan to lay a foundation for the well-being of the local economy in general. The *Legislative Guidebook* recommends coordinating local efforts with regional and state objectives as well as promoting local strengths and resources. Energy enters the economic development picture in two ways. First, diligent efforts to increase energy efficiency can have positive impact on the bottom line of any business, and second, both renewable energy and energy efficiency are beginning to open incredible economic opportunities for the future.

Economic Development and Energy

When a community has established a net zero energy or net zero emissions goal as part of its general sustainability goals, the energy plan for its commercial and industrial sectors must be as lean and clean as plans for the residential sector or for municipal facilities. In the area of economic development, investments in efficiency will reduce the carbon footprint of the business sector and strengthen its bottom line as well. Energy efficiency needs to be top priority for existing businesses as well as for any new commercial construction planned as part of future growth. As for renewable energy sources, they not only replace fossil fuel power, but their development represents potential local business opportunities.

Energy Efficiency Is Good Business

Improving the energy efficiency of commercial buildings is a sound strategy in stabilizing the local economy, but it is a difficult challenge. Energy efficiency has never been high on the list of priorities for commercial building developers. These buildings are generally built for short-term speculative investment purposes rather than long-term occupant satisfaction. Developers provide enough amenities to attract tenants but nothing to help them contain energy costs as time goes on. Building owners or managers don't pay the energy bills so they have little reason to demand efficient building construction. It is this energy disconnect between developer, owner, and tenant that makes the adoption and enforcement of commercial building energy codes like ASHRAE 90.1 necessary.

No one but the tenant has any financial interest in improving the performance of the building. However, if the municipal government has established energy efficiency goals for the whole community, it assumes an interest in the energy performance of the local commercial building stock and is in the position to expect compliance with the codes. Local government can also create incentive-based, voluntary collaborations with developers, similar to ENERGY STAR for home builders that raise the bar even higher. As we build new office parks and other commercial buildings, we can make significant progress in cutting energy use by requiring them to meet high performance standards. Communities can begin requiring "big box" stores to occupy less land or to redevelop abandoned properties by employing underground parking or building more than one story. Eventually, high-performance building construction will become the standard for commercial buildings. However, for the reasons stated above, this will not happen without direct intervention of local government in adopting and enforcing an energy code, and in encouraging developers to raise their own standards even higher.

Utility bills can represent a significant expense to a small business, and energy efficiency is a way of cutting costs. Savings can be found in improvements to the building where the business is located, or in reducing the "plug load," which includes computers, copiers, and other office equipment. If businesses lease their space, they have no control over insulation levels in the walls and roof or the efficiency of the windows. Their options are often limited to installing more efficient lighting or purchasing ENERGY STAR–qualified equipment. However, if energy efficiency is part of a community-wide economic development plan, commercial building owners and renters can be encouraged to pool their resources and work with the local government to achieve these goals.

Energy Efficiency and Downtown Revitalization

In downtown business districts where foot traffic has left for the mall on the edge of town or for more glamorous shopping areas elsewhere, the struggle for survival could be lost to a hike in the electric rate. Municipalities wanting to revitalize a decaying commercial district can make it a priority to increase the energy efficiency of the area, in both existing and new buildings, and can thereby attract the active participation of building owners and bring in new investors and business tenants. For example, if a whole block of buildings were being renovated by individual owners, these owners could split the cost of an energy audit for the whole block and contract for installation of efficiency measures, such as insulation and windows, from one contractor. Volume pricing and transportation savings could make the costs more affordable for everyone. Reduced energy use would save the businesses money and would also contribute to the net zero energy goals of the community. Their buildings would be more attractive to new tenants because of lower energy costs.

Another potential energy efficiency strategy for downtown areas is district heating and cooling. District steam heating used to be common in U.S. cities but now is primarily found on university campuses and large hospital complexes. Instead of having many small boilers, one large boiler serves a number of buildings within a limited radius. Efficiency of the system depends on the density of buildings being served. Now that higher urban densities are being explored as an antidote to sprawl and automobile-based lifestyles, the concept of district systems is reemerging, but with a few new twists. First, cooling systems are being added, using central chillers to circulate cold water. These are more efficient than the rooftop air conditioning units in common use now on most commercial buildings.

There are nonfossil sources of energy for district systems as well. One is waste heat from a generation plant or manufacturing facility. This application will depend on the proximity of the waste heat source to the ultimate users. It is not always economical to harness the heat from an existing plant. However, a combined heat and power (CHP) system is built specifically to use the heat generated along with the electricity, and these systems can be located to serve both purposes. Ground source heat pumps also work well for district energy systems, providing both heating and cooling, and could be powered by renewably generated electricity. One large field of underground piping could serve a number of buildings or a whole district.

These solutions don't necessarily pencil out to be economical in every situation. As with other efficiency strategies, district systems are generally more economical for new developments. A major expense for district systems is laying the pipe to serve the area. If a major renovation includes tearing up the streets, piping would be less expensive. The possibilities for district systems are still worth pursuing, particularly in conjunction with planning for high-density, mixed-use development. They offer a way to address residential and commercial energy efficiency simultaneously at the community level. District systems or CHP facilities might also work as part of a green office or industrial park. Creating an environmentally friendly, energy-efficient business climate is a strong marketing strategy in today's carbon conscious economy.

New Energy Businesses

All this emphasis on energy planning and reducing the carbon footprint will have a positive impact on economic development because it presents a number of exciting new business opportunities. As demand increases for energy-efficient products and installation of efficiency upgrades like insulation and new windows, new or expanded local businesses can be there to fill these needs. People will be looking for energy auditors to assess their properties and identify the most effective measures to be installed. Heating and cooling equipment suppliers and contractors will need to expand their expertise and product lines, and builders will be seeking new skills and information as the community demands higher energy standards.

The renewable energy market will grow as well, presenting additional opportunities. Municipalities may align with state economic development efforts that are focused on renewable energy resources such as wind or biomass. Sometimes it is just a question of specialized training and access to new suppliers. An existing plumbing firm can add solar thermal systems to its repertoire with little additional financial investment. Licensed electricians can become certified to install solar electric systems, and HVAC contractors can receive training for installation of ground source heat pumps. By courting these new markets, existing businesses keep up with changing times and strengthen the local economy.

More complex renewable energy enterprises such as wind farm development or manufacturing, producing biofuels, or making solar panels are economic development plums for a couple of reasons. Renewable energy industries are just beginning to establish themselves, and projected growth in these technologies is tremendous. This is the time to get in on the ground floor of what is recognized internationally as one of the strongest economic growth areas of the future. However, as international as this economic boom is predicted to be, the local nature of renewable energy means that much of the economic opportunity generated will remain local. In other words, a biofuels plant needs to be located near the cornfields or forests to remain economical, and a wind turbine developer that continues to maintain its wind farms will need local technicians to perform the work. For municipalities, the goal of attracting renewable energy companies integrates well into their smart growth economic development plan. If the plan also emphasizes effective use of energy, the community will share values with these prospective investors, offering an attractive incentive to bring this new business to town.

Natural Hazards

According to the *Legislative Guidebook*, the Natural Hazards element anticipates and lays out strategies for dealing with potential destructive acts of Mother Nature such as flooding, tornadoes, hurricanes. or wild fires. Whether we like it or not, anticipating extreme weather events has become part of smart growth or sustainability planning. We are already experiencing some extreme weather patterns and events, and scientists predict there will be more as the earth's atmosphere gets warmer. There are many areas of concern in natural hazards planning, from immediate response to long-term recovery, addressing everything from transportation to shelter, food, and communications, as well as longer term environmental and social recovery.

Energy Security

Among energy professionals, disaster planning usually falls under the category of "energy security," which includes anticipating a disruption of the energy infrastructure that is either natural or human-made. The phrase "energy security" might bring visions of grim federal agents and state-of-the-art surveillance equipment positioned around a nuclear power plant or hydroelectric dam. The threat to our energy infrastructure from terrorism is a very serious concern, particularly for large or strategically placed facilities. However, for most communities in the United States, the chances are far greater that the threat of disaster will come from natural forces.

Hurricane Katrina took us all by surprise. There were a lot of complaints about the insufficient and slow response of the federal government in providing food, shelter, fresh water, and energy to New Orleans and stricken areas along the Gulf Coast. Even though much of the criticism was well deserved, it was unrealistic to expect the federal government to mobilize itself from Washington, D.C., as a rapid response team. Because of its size and pace, the federal government serves best in the long term. In the case of New Orleans, this might be engineering preventive water control measures, including reestablishment of natural estuaries, or help with the gradual rebuilding of the city's neighborhoods. A more appropriate federal role in assisting the short-term disaster response scenario might be providing funding and coordination assistance to states and municipalities so they can make their own disaster mitigation plans before anything happens.

Neither Louisiana nor New Orleans was prepared for the results of a weather event of such unprecedented and dramatic proportions. Hurricane Katrina has become the clanging alarm and the symbol of global warming wrath that reminds us of our vulnerability. Hurricanes may be limited to coastal areas, but inland there are tornados, flooding, mudslides, and wildfires, each potentially devastating in its own way. While the potential for terrorist activity remains, we are much more likely to be hit with unrelenting weather. In short, global climate change will always be a very local experience, and communities need to take responsibility for their own disaster preparedness.

The energy portion of the Natural Hazards element needs to focus on providing reliable power to essential services in the community both during the weather or climate event and in its aftermath. Electricity is essential for a community responding to a weather emergency, and yet the transmission and distribution systems are themselves vulnerable in these situations. We remember the stories about the police and fire crews in New Orleans who couldn't communicate with each other because they couldn't recharge their cell phones and radios.

Energy security planning can be smoothly integrated into an energy-conscious smart growth plan. The initial plan could include distributed generation technologies, which are small generation systems located at the point of use. There might be electricity generated by the methane from an anaerobic digester at the sewage treatment plant or by solar electric panels on the senior center and a wind turbine at the local industrial park. Perhaps the hospital has installed hydrogen fuel cells as backup. During normal times, these generators would provide power at their locations, or perhaps sell excess back to the local grid if they are net metered.

The energy security plan would create levels of priority and put into place a system of electronic controls to redirect power according to those priorities in the event of an emergency. Primary services like police, fire, and hospital facilities receive highest priority, then other users are assigned priority levels depending on how essential they are. If the hospital has its own backup system, then power generated by the sewage treatment plant and the solar panels could be directed to the local emergency grid to operate police and fire communications equipment. Evacuation routes would need functioning street lights and traffic signals. The security plan would be set up to anticipate potential difficulties and points of greatest need. Other mitigation strategies, such as emergency shelter and food supplies could be integrated into the same plan.

An energy self-reliant municipality could remain connected to the regional transmission grid, giving and taking power, but be assured of backup in case of a local emergency. However, because of the presence of all that reliable local generation, it would also be possible to design a temporary disconnect system in case of a rolling black-out or disruption on the regional grid. Net zero energy planning, meshed with an energy security plan, would accomplish energy self-reliance without community isolation.

Outdoor Resources: Critical and Sensitive Areas and Agricultural, Forest and Scenic Preservation

These two elements cover planning for the conservation and management of all outdoor resources that are located within the municipality's jurisdiction. Critical and sensitive areas include forests, endangered species, and wildlife habitat, along with groundwater, stream corridors, floodplains, and wetlands. Agricultural and forest land resources are those in use by the community including productive agricultural land, parks and open spaces, and cultural and recreational areas. The primary connection of both these elements to community energy planning is indirect but undeniable. Natural habitat reaps significant benefits from cleaner air and water. If we reduce fossil fuel use we will reduce mercury, sulfur dioxide, and nitrogen oxide pollution as well as ozone and the other elements that make up smog. Although we have made progress in reducing acid rain, we have not yet eliminated the negative impacts of fossil fuel combustion on our natural environment.

There are also three specific energy issues that fit within one or both of these elements. First is the management of forest and agricultural land for biomass energy resources. Second is addressing the issues emerging around rural renewable energy, particularly manure management and wind farm and turbine siting. The third is management of parks and recreational areas to derive maximum energy benefits.

Biomass Fuels

Rural and agricultural communities are beginning to feel the heat from the bioenergy revolution. There is tremendous pressure from the federal government in particular to find something besides Middle Eastern oil to put in our gas tanks. Agricultural states have responded with patriotic and competitive zeal. Production of ethanol, the most established biofuel, has dramatically driven up total acreage planted in corn. Biodiesel, which can be made from a variety of feed stocks, including crop and wood waste, is being developed for widespread commercial production. Meanwhile, electric utilities, seeking ways to reduce their CO_2 emissions, are looking at biomass as something to burn with coal in order to accomplish this. They are considering wood waste and switchgrass as potential candidates.

Farmers may have land in the Conservation Reserve Program (CRP), a U.S. Department of Agriculture program to remove from production land prone to erosion or in other ways environmentally sensitive. They receive rent payments when the land is planted in grasses and other soil retaining plants. In this way erosion and toxic runoff are reduced, wildlife habitat is created, and farmers receive payment for participating in this conservation strategy. With new pressures on farmers to grow more corn, CRP lands are in danger of being returned to production as the rental agreements end, even though there is research in progress to determine whether planting these lands in perennial "biomass" grasses might be a better energy investment that would also retain the environmental advantages.

Forest lands are also under pressure as bioenergy becomes an economic opportunity. Management practices have improved dramatically since the mid-nineteenth century when the forests of the upper Midwest were leveled to build Chicago. Sustainable forest management has taken hold in many areas, and wood harvested from these forests holds extra value for builders seeking a LEED certification for a project. The challenge with forest land is that the overall ownership structure is changing. In the past, large parcels of land were the norm. The owner of several thousand acres would make the management decisions, for better or for worse, that would control the entire area. In recent years, these large holdings have frequently been broken up into many small parcels in the twenty- to forty-acre size range, often purchased by private individuals for their own recreational use rather than as a resource investment. If a timber company or a wood pellet manufacturer wants to harvest wood or collect wood waste in the area, they frequently have to sign contracts with several owners. Forestry officials are concerned that management of formerly large parcels is becoming uneven and sustainable harvesting more difficult and expensive to manage.

Towns that depend on agricultural or forest land for their livelihoods may not have direct planning jurisdiction over those lands. However, it is in the interests of the community to recognize the value of the biomass resources in their area as part of their energy planning, and to include goals to encourage cooperation among landowners to develop these resources in ways that benefit everyone.

Rural Renewable Energy

For rural municipalities, there are two other renewable energy sources worth assessing. One is animal manure and the other is the wind. Large livestock farms or dairies might find it economical to install an anaerobic digester to process animal waste, separating out the methane gas and organic solids as products that can be used or sold. Manure management is regulated at the state level, but local governments are the ones caught in the crossfire when nonfarming neighbors object to odors or the potential for pollution of waterways. If treated as a source of energy and fertilizer, manure can be a community asset rather than a hazard.

We have addressed wind energy thoroughly in another chapter, so it is simply noted here as another important energy resource to be assessed in a community energy plan. Because wind turbine installation has been controversial in some parts of the country, it is a good idea for a municipality to be informed and prepared in advance. Communities can easily have the local wind resource assessed to see if it is likely to attract wind farm developers and plan accordingly. Wind speed at a site is usually measured over a period of time using anemometers mounted on towers at different heights. There are also state wind maps available that show the general potential. The American Wind Energy Association (AWEA) offers a model ordinance for siting small wind turbines.

Parks and Energy

There are many good reasons to plant more trees, and now we can add to the list their ability to store carbon. Any plan for managing parks and forested recreational land should include tree planting or reforestation as a carbon sequestration strategy. It may eventually be possible to earn carbon credits for doing so. Park facilities can also contribute to net zero energy status by using renewable energy. Examples include solar water heating for pools and changing-room showers, day-lit restrooms, and solar electricity for remote lighting and communications.

Green Roofs

These elements cover all outdoor natural areas and habitats, so let's mention an innovation that creates new natural habitat in urban areas—green roofs. Chicago has embraced the green roof concept as its solution to calming the heat island effect in downtown areas. There are already over 200 green roofs installed on Chicago buildings. Heat islands are areas of the city where there is so much asphalt and concrete to reflect heat from the sun that the ambient air temperature rises significantly, causing the air conditioning load to rise as well. Green roofs use plantings of low maintenance vegetation in a special lightweight growing medium that covers whole sections of roof area. They can absorb heat and help manage storm water by soaking it up and using it. A green roof can earn a LEED credit for these energy efficiency and storm water management functions. Depending on the style installed, green roofs can serve these simple functions or they can be designed as more elaborate urban wildlife habitat or pleasant gardens for the building's occupants. Adopting the green roof concept is one way a community can extend the natural environment into its urban center.

Program of Implementation

The *Legislative Guidebook* makes a point of including a mandatory implementation element in its recommendations. This is particularly important for meeting a net zero energy or energy self-reliance goal as part of a smart growth plan, because energy enters into almost every element. Many specific suggestions have been offered already for implementing energy components of a plan. We've discussed energy efficiency building codes, renewable energy–friendly subdivision ordinances, ordinances for siting of renewable energy generation, and energy security planning. We've also mentioned goals for collaboration among government officials, businesses, and utilities, as well as the education of citizens and municipal staff about the energy goals. This element is really about organizing all the elements of the plan in order to achieve the overall goals and vision presented in the Issues and Opportunities element.

For the energy components, a realistic timeline, a designation of roles and responsibilities, and the creation of benchmarks to measure the progress made in carrying out the plan will be essential to success. The third requirement, to create a mechanism for measuring progress, will need to include establishing the energy baseline for the community so there is something to measure against. The data gathering on current use, opportunities for efficiency and clean energy applications, and projections of population and economic growth will be vital to making a realistic plan.

Something else that applies specifically to energy planning is systematically analyzing the costs and benefits of the many options available for improving the community's effectiveness in using energy. It might be economical to heat the high school with a wood-fired boiler, but marginal to build an anaerobic digester at the sewage treatment plant. Every community has different energy resources, priorities, and needs, which is why the planning process is so important. We need to understand how much renewable energy will be required to replace the fossil fuels we use, how reducing our demand will affect that requirement, and how we can economically arrive at the best balance between the two. Technical and economic analysis of potential solutions forms the foundation of a workable community energy plan. However, like many other aspects of sustainability planning, one viewpoint is not enough to get the job done. We need to begin integrating specialties and collaborating across professional and political turf. It's not just about the technology, or the economics, or even the environment. It's about integrating all those aspects and more into a sustainable whole.

Intergovernmental Cooperation

The model statutes in the *Legislative Guidebook* refer often to coordination and cooperation with other local government entities, regional agreements, and state legislation and policy. Smart growth planning in general recognizes the interconnectedness of systems, and energy is a case in point. Energy issues are currently on everyone's agenda even if only because of rising fuel costs, and officials at every level of government are becoming concerned about carbon emissions reduction as well. Because the laws of physics ignore political boundaries, we have begun to recognize where our traditional assumptions about jurisdiction may no longer be serving our best interests. An energy-conscious smart growth plan provides the ideal opportunity to spell out objectives for collaborating on mutually agreeable energy efficiency and renewable energy goals, whether in schools, local transportation systems, or on utility and other infrastructure planning. Collaborative goals with municipal neighbors might also interest them in exploring energy planning in their own communities.

Utility planning frequently extends beyond town or county lines. Opening communication with neighboring local governments about energy planning helps everyone better understand regional issues and priorities. This prepares communities for working more effectively with utilities to develop workable agreements regarding the regional growth pattern of utility infrastructure.

Because many actions concerning carbon emissions reduction are initiated at the state and potentially the federal level, intergovernmental cooperation needs a broader interpretation for energy issues. Many states are establishing renewable energy portfolio standards, carbon emissions reductions goals, transportation efficiency programs, and other energy-related actions. Some offer financial incentives and other benefits to individuals and communities for participating. A sustainable energy plan at the community level needs to look at state level commitments, pledges, and standards in order to take advantage of both the momentum and the available resources, and to remain in tune with state goals. States will increasingly rely on local governments to assist in meeting state government carbon reduction goals.

NOTES

1. American Planning Association, *Growing Smart Legislative Guidebook: Model Statutes for Planning and the Management of Change*, 2002 edition, Stuart Meck, FAICP, gen. editor, downloaded December 21, 2007, from http://www.planning.org/growingsmart/.

2. U.S. Environmental Protection Agency, *Electricity from Municipal Solid Waste*, downloaded September 17, 2007, from http://www.epa.gov/cleanrgy/muni.htm.

RESOURCES

Sustainability Programs, Standards and Challenges

Architecture 2030
An independent nonprofit organization founded by architect Edward Mazria in 2002 in response to global warming. The organization's mission is to reduce energy use by the building sector dramatically, and it has issued the 2030 Challenge to the architecture and building design community, as well as to government officials at all levels, to pledge participation in a stepped effort toward carbon neutrality of all new buildings by 2030. http://www.architecture2030.org/home.html

ENERGY STAR Green Buildings
http://www.energystar.gov/index.cfm?c=green_buildings.green_buildings_index

The Natural Step
An international sustainability organization that has developed a sustainable design process for communities, with projects in the United States.
http://www.naturalstep.org

The Congress for New Urbanism (CNU)
Website of the original New Urbanist movement, founded in 1993 to promote walkable communities as an answer to sprawl.
http://www.cnu.org

New Urbanism
An independent nonprofit website promoting good urbanism, smart transportation, transit-oriented development, and sustainability.
http://www.newurbanism.org

U.S. Environmental Protection Agency (US EPA)
Green Buildings web page.
http://www.epa.gov/greenbuilding

U.S. Green Building Council (USGBC), LEED for Neighborhood Development (LEED-ND)
http://www.usgbc.org/DisplayPage.aspx?CMSPageID=148

Energy Security

Sandia National Labs
Toward an Energy Surety Future, Sandia Report SAND2005-6281, October 2005, is a technical but readable report on an energy security model for a sustainable energy infrastructure.
http://www.prod.sandia.gov/cgi-bin/techlib/access-control.pl/2005/056281.pdf

Green Roofs

Greenroofs.com
Green roof industry web resource.
http://www.greenroofs.com

Green Roofs for Healthy Cities
Website of the green roof industry association formed in 1999 to promote installation of green roofs.
http://www.greenroofs.org

Smart Growth

Smart Growth Online
The website of the Smart Growth Network, a consortium of organizations formed in 1996 when the U.S. Environmental Protection Agency joined with several nonprofit and government, including environmental, groups, historic preservation organizations, professional organizations, developers, real estate interests, and local and state government entities.
http://www.smartgrowth.org/default.asp

U.S. Environmental Protection Agency (US EPA)
Smart Growth Pages on the EPA website.
http://www.epa.gov/dced

Wisconsin Department of Administration
Wisconsin Smart Growth Law: Wisconsin's Comprehensive Planning Legislation
http://.www.doa.state.wi.us/dir/documents/CompPlanStats_0504.pdf

Sustainable Communities

Institute for Sustainable Communities
Nonprofit organization dedicated to helping communities become sustainable and focusing on collaboration and community building to address environmental and resource issues.
http://www.iscvt.org

Rocky Mountain Institute
RMI is an independent nonprofit organization that has promoted energy efficiency and sustainability since 1982. They have developed materials to assist communities in sustainable energy planning. These and other sustainability publications are available through their bookstore.
http://www.rmi.org

U.S. DOE, Smart Communities Network
The U.S. Department of Energy website for sustainable energy use in communities.
http://www.sustainable.doe.gov

Sustainable Communities Network
This organization as founded by CONCERN and the Community Sustainability Resource Institute in 1993. CONCERN, Inc., is a national nonprofit environmental education organization. Their website offers publications and links to other sustainability organizations.
http://www.sustainable.org

Another Helpful Website

Database of Incentives for Renewables and Efficiency (DSIRE)
This is an excellent source of information about financial incentives and other programs offered by individual states to promote renewable energy and energy efficiency. It is maintained by the North Carolina Solar Center in partnership with the Interstate Renewable Energy Council (IREC), funded by the U.S. Department of Energy. http://www.dsireusa.org

CHAPTER SIX

Clean Energy Policy
and the Government Role

Looking at the energy challenges in our future brings to mind the old Chinese curse, "May you live in interesting times." Considering how much energy we use, and the projected environmental impacts of continuing to use our trusty fossil fuels, moving toward the future will be nothing if not interesting. It is easy to panic or fall into despair about how we will accomplish the huge transition to clean alternatives before CO_2 levels in the atmosphere bring about the changes that can potentially overwhelm the world as we know it. We have never before faced a threat as comprehensive as global warming.

For some time, policy analysts, scientists, energy engineers, and other experts have been discussing whether we have the technical potential to cut carbon emissions sufficiently to avert serious climate change. While there are many different opinions about how it might be accomplished, most agree that we have the technological capability to slow global warming by moving to a combination of alternative fuels and clean energy resources along with radical improvements in energy efficiency. But calculating the square footage of solar panels or the number of wind turbines we will need, or projecting how many cars the present corn crop will fuel, is just collecting data.

The development of advanced energy technologies like hydrogen fuel cells or solar nano-technology appears to move us closer to our goal of using clean energy, but these too are, in a sense, just gathering information. We are learning what new technologies we are capable of building and how we can harness different energy resources. Even wind power, which has grown dramatically as an industry over the last five years, is still only installed at the scale of a technical demonstration in the United States. Those who make and sell wind turbines or solar panels or ethanol made from corn are working hard to build interest in their products, and they have made impressive progress. They are gathering sound evidence that these things work and that they can create new energy industries and professions. However, even this is not enough to launch a new energy age. We will need more than good data, advanced technology, and solid installation experience in order to make definitive steps toward reducing carbon emissions. We need to develop the kind of energy policy that makes a commitment to getting us there, rather than relying on policy dictated solely by market forces with a dollop of environmental regulation plopped on top for effect.

Reliance on market forces has driven energy policy in the past. Policy at the federal level has favored the fossil fuel industries because they could justifiably claim to be powering the nation. What was best for the railroads, the automakers, and the utilities seemed to be what was best for the country because it meant that our economy and prosperity would keep humming along. However, this somewhat myopic view failed to include any negative impacts. Energy economics within the industry confines itself primarily to determining the least-cost options for energy production, whether extracting, refining, or generating. No price tag is assigned to natural resources like clean air or water, which are therefore regarded as "externalities" because no one can say what they are worth. The environmental costs of using fossil fuels are typically absorbed by the public sector and paid for by our tax dollars. Up to the present this has been an acceptable trade-off for cheap energy. Now, however, there is compelling evidence that externalities will begin to include severe weather events that will likely increase as global warming progresses. What is it worth to us to prevent hurricanes like Katrina from bludgeoning the Gulf Coast every summer?

Governments Begin to Take Action

The situation clearly demands unprecedented levels of cooperation both internally and internationally. Recognizing this, we look to our government to organize our response and help us find answers. The pressure is on government at all levels to develop policy that will reduce the carbon footprint. The federal government has been slow to acknowledge the urgency of the situation and has trotted out some additional subsidies for development of advanced fossil fuel technologies and alternative fuels, but it has not drafted a substantive carbon reduction policy. The 2005 Energy Policy Act was a hodge-podge of subsidies and political trade-offs that once again confirmed heavy federal support for fossil fuels and nuclear energy and gave comparatively low recognition to renewable energy and energy efficiency technologies, presenting no cohesive approach to reduction of energy use or carbon emissions. The Energy Independence and Security Act of 2007 gave us higher fuel efficiency standards for vehicles, more money for developing biofuels, and some new building and appliance efficiency standards, but Congress was unable to agree on the value of a national Renewable Energy Portfolio Standard for utilities or to renew the Production Tax Credit for wind power.

The federal government has seldom been on the cutting edge with regard to innovative policy or legislation on controversial issues. It has often waited until individual states have hammered out the details on effective laws or programs. Over the years the states have originated a variety of environmental laws that eventually led to federal legislation, and the same can be said of occupational safety and energy legislation. However, state governments have limited resources for leading the charge, and eventually individual state efforts can create a national patchwork of laws and regulations that becomes economically and politically inefficient. With regard to climate change and carbon reduction, a commitment at the federal level will ultimately be essential to making headway, but it will come as the result of a groundswell of public opinion as state and local governments hammer out strategies, goals, and action networks.

State and Local Efforts

Over the last ten years there has been considerable policy and planning activity at the state and local government levels. This is hardly surprising since the actual reduction of carbon emissions happens locally at millions of homes, businesses, and institutions across the country, and it will be locally that economic and environmental impacts of climate change will be felt first. State and local governments have begun to grasp the importance of being proactive about setting carbon reduction goals and making them happen, recognizing that both market forces and federal action, left to themselves, move too slowly.

California has been particularly successful in breaking new ground. To reduce vehicle emissions, it passed the California Vehicle Global Warming Law in 2002, which requires a 30 percent reduction in auto emissions by 2016. So far, twelve other states have adopted similar legislation despite lawsuits by auto makers, and four additional states are planning to do so. California has opened the door for higher efficiency vehicles to enter the U.S. market by using its consumer power to add muscle to the law. Every state that subsequently joins California in its requirements simply increases the market and availability of these models. The Bush administration has not taken kindly to this state level approach to emissions reduction. In December 2007, the U.S. Environmental Protection Agency took steps to block California from proceeding with its new standards, claiming that new federal fuel efficiency standards are sufficient. California's Governor Schwarzenegger will sue the agency in what promises to be a landmark confrontation in the arena of federal versus state authority to pass environmental legislation.

State governments have taken a number of approaches to policy that would reduce carbon emissions. Of necessity, they must make the best use of their limited resources to address a multiplicity of issues, and many have recognized that plans for reducing carbon emissions can also produce a cleaner environment, contribute to a more stable energy supply, and offer economic opportunities to local industry. Depending on a state's energy needs and resources, it might consider increasing generation of electricity from renewable energy, encouraging biomass or biofuels production, instituting a system of carbon credits, or mandating higher standards of energy efficiency in state facilities and buildings. Many governors have started by convening a task force or fact-finding commission to define needs, resources, and goals and to assess potential impacts of regional or national trends within their own states. Task force appointees usually represent a cross-section of interests and areas of expertise, from academia, the nonprofit sector, and the environmental and business communities.

Some task forces address specific solutions, such as New Jersey's Carbon Dioxide Reduction from Trees Task Force, Washington's Commute Trip Reduction Task Force, or The Governor's Allocation Task Force in Oregon that is charged with developing an emissions allowance standard. Other states are taking a closer look at specific environmental or economic impacts in store for their residents. The Illinois Task Force on Global Climate Change began in the early 1990s to look at the future economic implications of its vast coal resources, and Hawaii's Greenhouse Gas Emissions Reduction Task Force is addressing potential effects of sea level changes on its tourist industry. Several states, including Wisconsin, Minnesota, and New York, have also convened task forces to address climate change and carbon emissions from a broader perspective, considering both emission reduction strategies and opportunities for development of clean energy resources within their states.

These task forces are a first step in determining what actions need to be taken and whether these actions will be most effective as carrots or sticks. Typically, task forces are asked to recommend goals for carbon reduction or clean energy use, and they will often commission a study of the state's carbon footprint or of the role of energy in its economy. At this writing, at least sixteen states have set CO_2 emission reduction targets, with long-term goals averaging between 50 and 80 percent below 1990 levels by 2050. The next step in the process is for the governor and the legislature of a state to take the task force reports and recommendations and draft legislation that will accomplish these goals.

Examples of State Emission Reduction Targets

California was the first state to pass comprehensive legislation making greenhouse gas reduction goals mandatory. In September 2006, Governor Arnold Schwarzenegger signed into law AB32, the Global Warming Solutions Act, which makes use of both regulatory and market mechanisms to help California reach its CO_2 emissions goal of 80 percent below 1990 levels by 2050. Hawaii, New Jersey, and Connecticut have passed legislation as well, and other states are considering similar bills. On the regulatory front, state commissions in twenty-four states and the District of Columbia have established various Renewable Portfolio Standards (RPS) requiring electric utilities to acquire established minimum levels of renewable energy in their generation portfolios. While these standards are not specifically aimed at reducing greenhouse gas emissions, they will contribute to the effort. The failed Renewable Portfolio Standard introduced into the 2007 federal energy legislation will continue to garner support based on political pressure from state-level efforts, and it is only a matter of time before it will pass.

Regional Efforts

States are also forming regional coalitions to deal with global warming issues. The Regional Greenhouse Gas Initiative (RGGI) brought together seven northeastern states with the goal of creating a regional carbon cap and trade program. A cap and trade program defines the maximum level of emissions that will be allowed, with specific limits assigned to utilities and other high-level emitters. These companies are responsible for either lowering their emissions or obtaining carbon credits from other companies that emit less than their limit. After a period of time, the cap is lowered, forcing companies to find new ways of reducing emissions or buy more carbon credits from other sources. The idea began with New York Governor George Pataki in 2003 when he invited neighboring states to address emissions from electric power plants in the region. The participants were aware that emissions from power plants know no borders, so instead of complaining about each other's traveling pollution, they decided that they would all be better off tackling the problem as a region.

After over two years of study and discussion, a plan emerged outlining the first regional mandatory cap and trade program in the United States. Participating states are Connecticut, Delaware, Maine, New Hampshire, New Jersey, New York, and Vermont, and Maryland is planning to join. In August 2006, the participants agreed on a model regulation that they would be take back to their own legislatures. According to the model, an emissions cap will be established in 2009 and then strategically lowered over time to achieve a 35 percent emissions reduction by 2020. Power plants and other greenhouse gas emitters will be required either to cut emissions or purchase carbon credits from those who are already below their cap limit. States will employ several strategies to meet their mandates, including renewable energy and clean coal technologies, energy efficiency, and ratepayer rebates.

Municipalities and Local Governments

Local governments are finding strength in numbers through organized networks that tie municipalities, counties, and other local government bodies together. One of the most well-known networks sprang from the U.S. Mayors Climate Protection Agreement, generated by Seattle Mayor Greg Nichols in February 2005. Launched in response to U.S. reluctance to ratify the Kyoto Protocol, the Agreement was adopted by the U.S. Conference of Mayors in 2005 and began drawing cities and towns from all over the country who wanted to sign on in support of definitive U.S. action on global warming. As of this writing, there were close to 670 mayoral signatures from towns and cities in all fifty states and Puerto Rico.

Table 2
Examples of State Emission Reduction Targets

State	Short Term	Medium Term	Long Term
Midwest			
Illinois			60% below 1990 by 2050
Iowa			50% below base-line (TBD by Council) by 2050
Minnesota	15% below 2005 levels by 2015	30% below 2005 levels by 2025	80% below 2005 levels by 2050
West			
Arizona		2000 levels by 2020	50% below 2000 by 2040
California	2000 levels by 2010	1990 levels by 2020	80% below 1990 by 2050
New Mexico	2000 by 2012	10% below 2000 by 2020	75% below 2000 by 2050
Oregon	Stabilize by 2010	10% below 1990 by 2020	75% below 1990 by 2050
Washington		1990 levels by 2020	50% below 1990 by 2050
North East			
Connecticut, Massachusetts, Main, New Hampshire, Rhode Island, Vermont	1990 by 2010	10% below 1990 by 2020	75–80% below (baseline varies from 1990 to 2003)
New Jersey		1990 levels by 2020	80% below 2006 by 2050
New York	5% below 1990 by 2010	10% below 1990 by 2020	

Table reproduced courtesy of the Wisconsin Department of Natural Resources

The agreement recognizes the U.S. contribution to global warming and the emissions target of 7 percent below 1990 levels by 2012 that the Kyoto Protocol would have imposed had the United States ratified it. The agreement also notes the many local government and corporate emission reduction efforts already underway nationally. By signing the document, mayors join in urging the federal government to take substantive action on global warming to meet the Kyoto mandate, including reduction of dependence on fossil fuels, by increasing efficiency and use of renewable energy resources. But more important, mayors pledge to address emission reduction in their own communities by first taking inventory of energy needs and resources and then setting specific goals for taking action. The agreement lays out several areas to address such as applying land use policies that reduce sprawl, investigating public transportation options, developing local renewable energy resources and buying "green tags," using the local code structure to raise efficiency levels, and improving efficiency in municipal buildings and facilities. Many more strategies are mentioned, from buying ENERGY STAR products and adopting the LEED® rating system to waste handling and recycling, urban forestry, and alternative fuels for municipal fleets. The U.S. Conference of Mayors established the Mayor's Climate Protection Center in 2007 to offer support and coordination.

International Opportunities for Local Action

Some U.S. cities participate in the international arena as well. In 2005, the Mayor of London, England, convened the Large Cities Climate Leadership Group, an international effort to bring together the mayors of the world's largest cities to address the fact that 75 percent of carbon emissions come from urban areas. A year later this group formed a partnership with the Clinton Climate Initiative, a Clinton Foundation project that has pledged to assist these cities in reducing energy use and greenhouse gas emissions by providing advice on management of volume equipment purchases, technical assistance, and standard tools to track and measure results. The goal is to establish these cities as leaders in carbon emissions reduction whose methods and results can be replicated among smaller cities. The twenty-two major cities that have joined the effort to date represent a remarkable cross-section of global culture. They are Berlin, Buenos Aires, Cairo, Caracas, Chicago, Delhi, Dhaka, Istanbul,

Johannesburg, London, Los Angeles, Madrid, Melbourne, Mexico City, New York, Paris, Philadelphia, Rome, Sao Paulo, Seoul, Toronto, and Warsaw.

Five Basic Assumptions about Dealing with Global Warming

The world is abuzz about finding solutions to the global warming crisis. We are now in the stage of accepting the immensity of the problem and we are deeply immersed in discussing what to do next. The purpose of any level of government policy is to get ourselves organized, as can be seen in the examples described above. Primary to any policy is the central goal—in this case, it would be reducing greenhouse gas emissions. Whatever detailed action is spelled out in the policy, and whatever tools are employed, they all serve that central goal. The goal itself, however, is always based on political, economic, or technological assumptions. Global warming is of a magnitude that is unique in our experience because it involves everyone on the planet. Our policy assumptions will need broadening to incorporate the complexity and scale of the problem. To get things started, here are five suggested assumptions for those formulating policy about the global warming challenge.

1. There Are No Magic Bullets

Coal, oil, and natural gas are something like magic bullets themselves because they each do so many things. Using fossil fuels has given us the idea that magic bullets exist, at least as far as energy is concerned. However, moving beyond fossil fuels will require a willingness to try many strategies. We can't simply replace coal, oil, or natural gas with a renewable substitute and forge ahead. First, we need to reduce the amount of energy we need to get things done. Second, renewable energy and fuels have their own limits of locality, which means that the strategies we choose will depend somewhat on geography and local resources. And third, economic priorities vary from place to place, within the U.S. and among nations internationally.

A successful climate change policy will need many components, each contributing to the whole solution. Princeton professors Robert H. Socolow and Stephen W. Pacala have come up with a concept of "wedges" to narrow the space between the stabilized emissions path we would like to achieve and the sharply rising emissions scenario that is likely if we do nothing.[1] They have divided this difference into seven wedges or levels, each representing 25 billion tons of carbon emissions over the next fifty years. They also provide us with fifteen potential strategies that would each accomplish a wedge-worth of carbon emissions reduction. These are drawn from five different areas of energy technology: end-user efficiency and conservation, agriculture and forestry, renewable energy, carbon capture and sequestration, and coal-fired power generation. For example, under end-use efficiency and conservation, one wedge, or 25 billion tons of carbon emissions could be avoided if we were to "cut electricity use in homes, offices and stores by 25 percent." Another wedge of reductions would be accomplished if we were to "install carbon capture and storage at 800 large coal-fired power plants," or "expand conservation tillage to 100 percent of cropland." Some strategies improve on existing technology, some expand or develop new technologies, some reduce energy use or simply sequester more carbon naturally. According to Socolow and Pacula, only seven wedges will do the job for the planet, which means that individual nations and regions will have considerable choice about how they accomplish their share of the reductions.

The wedge concept has helped us wrap our minds around the problem and provides a helpful structure for global warming policy. We could divide the wedges differently, or establish additional strategies to equal the same results. Regardless, this concept keeps us on track and helps us remember that we can't sit around waiting for one big answer. It will be technological, economic, and political efforts in increments that will pay off.

2. We Have the Technology to Start Right Away

Technology does not come into common use simply because it works. If you build a better mousetrap, the world will not beat a path to your door unless mice have become such a problem that the traditional methods for catching them have failed. In other words, we can't assume that mainstream energy technologies, whether for transportation, generation, or various end-uses, have reached some pinnacle of performance or else we wouldn't be using them, and if other technologies really worked, we would be using them instead. There are always many options, and they await a shift in our priorities.

Over the last thirty years or so, great progress has been made in improving the energy efficiency of electrical appliances and equipment. However, it has been due only to programs like ENERGY STAR and other promotions that these products gained popularity and subsequent economic success. Of course, they work, but they also work for less cost over time. When the latter characteristic became a priority, these products began to gain market share, and volume of sales reduces the price. The same can be said for solar water heaters, which have outgrown their awkward stage and now perform admirably in many climates. They also work for less cost and help reduce carbon emissions as well.

There are many technologies already in existence that can help achieve wedges of carbon reduction, whether efficiency strategies, renewable energy, or sequestration technologies. When Sokolow and Pacala developed their wedge strategies, they counted only technologies that have already been commercialized somewhere in the world. Naturally, we will not stop inventing and developing new ones, but the point is we have plenty of expertise to get started.

3. It Will Take International Cooperation on an Unprecedented Level

The politics and economics of energy are increasingly in play internationally, and changes cannot simply be technological if there is to be true sustainability for everyone. More than ever before in history, the nations of the world are interconnected. We have reached the end of that frontier era where individual governments can divorce themselves from the world community, even though a few are still trying. There are simply too many of us now, elbow to elbow, to make anything other than cooperation a practical policy.

There are no nations that can currently claim total energy independence, and because we share the same atmosphere, we share the impacts of its degradation as well. Because every single nation has a stake in carbon reduction, we need a plan the whole world can agree to carry out. It must take into account the varying social, political, and economic circumstances of different countries and regions, and yet it must effectively do the overall job of reducing carbon emissions for everyone in a way that will continue to work as time progresses. Perhaps there will be a number of plans that intersect and interact to achieve the same goal. Whatever its form, the creation of international energy policy that reflects the enlightened self-interest of all nations to steer a course away from potential environmental disaster will require more effort and innovation than any advanced super technology we can imagine.

4. We Will Be Working for the Common Good

During World Wars I and II, Americans pulled together for victory by collecting scrap metal, buying war bonds, and planting victory gardens. Personal or partisan agendas didn't completely disappear, but most were put on hold for the duration. Social pressure and awareness of the gravity of the situation reminded us we were all in it together. When Jimmy Carter called the challenge of reducing energy the "moral equivalent of war," many thought him overly dramatic. He was quite possibly recalling his own experience of the national mindset during World War II when the United States worked together as a nation to do its part in defeating the forces of fascism and oppression. We took action and made sacrifices for the common good. There are still many people alive who remember World War II or recall the power of a local disaster to mobilize our better instincts. Their stories offer sufficient proof that we are capable of rising to such an occasion.

Global warming is even more compelling a problem than international fascism and oppression. As we hammer out policies, goals, and plans we will be called upon to transcend partisan, regional, and economic politics because with something as all-encompassing as global warming, our personal interests are intimately tied to the common good. Making decisions about land development, habitat, and resources will mean careful consideration of potential consequences not only for ourselves but for everyone. Doing this successfully will require us to both obtain good information and maintain open minds when we use it.

A good example is the issue of raising the fuel economy standards for vehicles. There are plenty of data to support the argument that raising the CAFÉ standards can significantly reduce our need for gasoline. This is a primary goal for reducing both carbon emissions and dependence on foreign oil. However, it was a real battle to craft new vehicle efficiency standards to be an average of 35 miles per gallon by 2020 as part of the Energy Independence and Security Act of 2007, hardly an impressive goal considering that Toyota and Honda hybrid cars are already achieving numbers in the 50 mpg range. Many legislators continue to cling to the auto industry's point of view that raising the fuel economy standards will cause national economic disruption. They say that the price of cars, SUVs, and pickup trucks will rise too high, or people won't want to buy efficient cars, causing a drop in demand and therefore loss of jobs, particularly in Detroit and other cities that produce vehicles.

Instead of accepting these traditional economic assumptions, let us consider other potential outcomes. First, economies of scale, brought into play by the requirement that all vehicles conform to the new standard, may mean that the price of an individual vehicle might not rise enough to cause economic disaster. Also, the high price of gasoline is beginning to make efficient cars more appealing, particularly since new efficiency technologies have made cars safer. Second, if the price actually did rise enough to cause a shift, perhaps the demand for public transportation options would also rise, creating more choices for everyone and saving even more carbon emissions. The losers are not necessarily the automotive workers either, who could be employed building buses and streetcars.

This scenario is offered only as an example of how things might change when policy is adopted for the ultimate common good. It is likely in everyone's interest eventually to reduce the number of cars on the road. We can either incorporate this idea into our policy planning, or we can wait until high gas prices cause our traffic and our economy to grind to a halt. Economic decisions will be the most difficult, but delaying investment in preventive measures will only increase costs down the line.

5. Life Is Going to Be a Lot Different

We know that global warming will be changing a lot of things in the future. Obviously our lives will change as well. This is not necessarily a bad thing. We can prepare ourselves to use different resources as we learn that traditional ones no longer serve us, thereby making things a lot easier. We need to take a longer view, anticipating market trends rather than waiting for the market to move itself. We don't need to continue to do things the same way. What we have has come about through a series of political occurrences, economic decisions, technological and geological discoveries, and our energy infrastructure will continue to evolve as our political will changes. We can choose to do something different, and our strength lies in our vast knowledge of energy systems and their environmental impacts, and our ability to communicate instantly with the rest of the world to share knowledge and energy solutions. What is new is the idea that we can choose to lay out a plan, and integrate with other nations. Energy issues are shared challenges. We have the opportunity to move forward as a community of nations by working together on energy.

Conclusion: It's Time to Get Real

Living sustainably means balancing our needs and resources so that we never run out of food or water, or the means to clothe or shelter ourselves. It also means that everyone has access to the system and is able to live this way. Sustainability is the ultimate perpetual motion, "quality of life" fantasy. It is a great theory that we can take what the planet provides and support billions of people indefinitely in gentle prosperity. However, even as we plan and dream of making this a reality, we know that life is lumpy and complicated, and we will need a lot more than a good theory. The truth is that we have never had anything even approaching sustainability that could be replicated globally at the scale we would need it today. Furthermore, our existing lifestyle, which creeps inexorably toward a high-tech urbanity, has a relentless momentum of its own.

How do we turn our direction toward what appears to be an impossible goal? We can begin by forgiving ourselves for living unsustainably in the past, so that we can roll up our sleeves and get to work. The trade-offs in our previous resource decisions were made based on much less awareness than we now possess, whether we are talking about creating pollution or promoting inequitable economic policies. If we can accept the fact that every environmentally unfriendly thing we currently do seemed like a good idea at the time, we can also acknowledge that it may be time to abandon concepts and technologies that are no longer beneficial. The new information and understanding we have gives us the freedom to make radical new choices, potentially moving us closer to true sustainability.

NOTES

1. Robert H. Socolow and Stephen W. Pacala, *A Plan to Keep Carbon in Check, Scientific American* (Sept. 2006), pp. 50–57.

RESOURCES

Regional Greenhouse Gas Initiative (RGGI)
A collaborative of Northeastern and Mid-Atlantic states to develop a regional strategy for controlling carbon dioxide emissions. A central strategy is the implementation of a multistate cap-and-trade program with a market-based emissions trading system.
http://www.rggi.org/

California's Vehicle Global Warming Law: California Clean Cars Campaign
http://www.calcleancars.org/About.html

U.S. Conference of Mayors Climate Protection Center
Launched in 2007 to provide guidance and assistance to mayors who have signed the U.S. Conference of Mayors Climate Protection Agreement and are leading efforts to reduce greenhouse gas emissions in their municipalities.
http://www.usmayors.org/climateprotection

APPENDIX A: Energy Connections to Sustainability Goals

A concise table showing examples of how energy issues intersect with sustainability goals such as environmental quality, diversified and efficient modes of transportation, eco-friendly structures, secure food systems and clean water resources, environmental justice, strong local economies, and healthy populations.

Sustainability Goals	Energy Connections and Current Issues
Environmental Quality: clean air, clean water, species diversity	• Pollution from coal-fired electricity and burning fossil fuels: NO_X, SO_2, mercury, CO_2 emissions • Lake and river impacts of industrial and power plant cooling • Multiple environmental impacts of damming rivers for hydropower • Habitat disturbance from transmission lines, roads, and sprawl development • Coal extraction: mountaintop removal and strip mining of coal with accompanying watershed degradation and pollution • Oil and gas drilling: site contamination, oil spills, habitat disturbance from oil and gas pipelines • Storage and transport of nuclear waste • Safe disposal of power plant combustion products and chemical plant wastes • Smog and ozone from petroleum-powered transportation • Virtually infinite legacy of plastics pollution
Diversified and efficient modes of transportation	• Dependence on petroleum, a rapidly diminishing and politically volatile fuel • Inefficient vehicles • Transportation inefficiencies inherent in sprawl communities • Energy embedded in continually expanding road and highway infrastructure • Continuing political priorities to invest in infrastructure for the private automobile

Sustainability Goals	Energy Connections and Current Issues
Eco-friendly structures, green building	• Construction and site: petroleum-based building materials, embedded energy in materials and equipment, heat island effect
	• Transportation of building materials and building users
	• Electricity and natural gas used for heating, cooling, and refrigeration systems, indoor and outdoor lighting, hot water, appliances, electronic equipment
	• Management of building waste stream: treatment of sewage, transportation and disposal or recycling of trash, gasoline to power equipment for removal of yard waste and snow from site
	• Ordinances and neighborhood covenants that outlaw or attempt to discourage energy efficiency or renewable energy technologies or practices
Secure food systems and clean water resources	• Diesel fuel for farm equipment to produce food
	• Petroleum-based fertilizers
	• Long-distance transportation of goods
	• Refrigeration and storage of food
	• Natural gas and electricity required for food processing and energy embedded in packaging
	• Impacts of climate change on local growing conditions
	• Disaster preparedness due to volatile weather patterns
	• Polluting runoff from streets, highways, and parking lots
	• Water supplies required for some desert coal mining techniques

Sustainability Goals	Energy Connections and Current Issues
Environmental justice	• Siting of power plants, substations, and nuclear waste repositories near underrepresented communities
	• Working conditions for coal and uranium miners and oil workers
	• Degradation of life quality around energy-extracting operations and refineries
	• Disruption of communities through environmental degradation near mountaintop removal operations
Strong local economy	• Energy utility payments leave the community
	• Inefficient buildings and operations contribute to making small businesses marginal and local government budgets tight
	• Development of renewable and efficiency resources creates local jobs
Healthy population	• Increased respiratory ailments from particulate pollution levels created by fossil fuel combustion
	• Genetic irregularities and other health problems from presence of petrochemicals in water supply
	• Obesity and poor fitness from car-based culture
	• Gradual poisoning by products made of carcinogenic plastics like PVC, made from petroleum byproducts

In every deliberation, we must consider the impact of our decisions on the next seven generations.

—The Iroquois Confederacy

This timeline is a sample of events, inventions, and discoveries that shaped the fossil fuel era in the United States. The development of coal, oil, and natural gas is shown alongside the growth of the electric industry, nonfossil fuel trends, and current historical events for reference. The main point of presenting it in timeline form is to show the trends over time rather than to include every milestone along the way. Many early inventions and experiments did not translate immediately to commercial success, but there has been tremendous momentum from the beginning toward growth and progress, two goals made possible primarily by fossil energy.

The adoption of fossil fuels as the primary sources of energy in the United States has happened since the American War of Independence. Most of what we regard as the age of fossil fuels has occurred during the past 175 years or so—not much more than the seven generations invoked by the Iroquois Confederacy in the meetings of their Council. How could we have known? These have been amazingly busy years, encompassing the concept of Manifest Destiny coupled with the resources and fuels to make it happen. Whatever can be said about the United States for better or worse, we are what we are because of fossil fuels.

Before 1775

Coal	Oil	Gas (Coal Gas and Natural Gas)	Electricity	Nonfossil Energy	Current Events
By the 1600s—Fuelwood shortages in England bring coal to the cities as the principal domestic heating and cooking fuel. Coal first began to replace wood and charcoal for domestic uses in Britain around 1000	1745—Oil lamps used on the streets of Paris	1626—Native Americans burning natural gas from seeps found around Lake Erie	1752—Benjamin Franklin discovers that lightning is electricity and invents the lightning rod	Before 1775: For lighting, people used tallow candles, or lamps burning animal fat, pine pitch, whale oil (very expensive), or turpentine/alcohol (very dangerous)	1607—Jamestown founded in Virginia. The following year the first exports include timber and iron ore
	1770—Oil street lamps erected in New York City	1670—Gas first distilled from coal as a byproduct of coke production and initially regarded as a novelty. Another byproduct of the process, coal tar, would eventually produce a wide variety of petrochemical products		Commercial and agricultural energy sources were animal and human muscle power, wind and water for milling, and wood, charcoal, or some coal for forging and other processes needing heat	1636—Harvard College is founded in Cambridge, Mass.
Early 1700s—Coke replaces charcoal in England as primary fuel used in blast furnaces for smelting iron				Residential energy sources were human and animal muscle power and fuel wood	1681—William Penn receives a Royal Charter from King Charles II to found Pennsylvania
1748—First coal mined commercially in America at Virginia				Transportation energy came from draft animals, poled barges, and other muscle-powered boating, and wind-powered sailing ships	1692—Salem Witch Trials, Massachusetts
1753—First steam engine arrives in America from England. Two years later it is employed to pump water from a mine					1701—Detroit is founded by the French
					1731—Benjamin Franklin establishes the first subscription library, in Philadelphia
					1754–63—French and Indian Wars

Before 1775 (continued)

Coal	Oil	Gas (Coal Gas and Natural Gas)	Electricity	Nonfossil Energy	Current Events
				1742—Benjamin Franklin invents the "Pennsylvania Fireplace" for home heating, now known as the Franklin stove. It is an open, cast iron stove resembling a fireplace but is safer and uses less wood	By 1760—Colonist population in America reaches 1.5 million
					1773—The Boston Tea Party
					1774—Meeting of the first Continental Congress
				1767—First solar cooker invented by Horase de Saussure, a Swiss naturalist. He cooked fruit in the oven, which reached 190 degrees Fahrenheit	1774—Englishman Thomas Priestly discovers oxygen

1775–1800

Coal	Oil	Gas (Coal Gas and Natural Gas)	Electricity	Nonfossil Energy	Current Events
1775—Coal field near Richmond, Virginia, is producing coal shipped to the eastern seaboard. By 1833, production reaches 142,000 tons per year		1792—Scottish engineer William Murdock begins pioneering work on coal gas lighting technology. He spends much of his career as an employee of James Watt	1800—Italian Alessandro Volta invents the battery	1783—First manned flight via hot air balloon, Paris, flown by Pilâtre de Rozier and the Marquis d'Arlandes. Fuel was likely to have been wool or straw	1776—Declaration of Independence
					1781—British surrender at Yorktown

1775–1800 (continued)

Coal	Oil	Gas (Coal Gas and Natural Gas)	Electricity	Nonfossil Energy	Current Events
1776—In England, Scottish inventor James Watt demonstrates the first efficient coal-powered steam engines, one to pump water from a mine and the other to operate foundry bellows			1800—Italian Alessandro Volta invents the battery	1787—American John Fitch holds the first demonstration of a steamboat, on the Delaware River, for members of the Constitutional Convention. He patents his invention in 1791	1783—Treaty of Paris ends war, grants American Independence
By the 1790s—Pittsburgh becomes first industrial center in the United States to use coal-fired steam power for manufacturing (glass and ironwork)				1793—Francois-Pierre Argand invented a lamp that burned whale oil or rape seed oil, also called colza	1788—U.S. Constitution Ratified
1794—The anthracite coal region of eastern Pennsylvania is purchased from the Iroquois Confederacy				1794—Eli Whitney patents mechanical cotton gin	1789—George Washington elected President and John Adams Vice President
					1789—French Revolution
					1790—Congress votes to locate Washington, D.C., along the Potomac on land donated by the colony of Maryland
					1790—The first U.S. census is taken. Non-native population is 4 million

1801–1850

Coal	Oil	Gas (Coal Gas and Natural Gas)	Electricity	Nonfossil Energy	Current Events
1812—The first commercially viable steam locomotives built for the Middleton Railroad in Leeds, England	1815—Oil extracted from brine wells in Pennsylvania is considered a nuisance	1806—First demonstration of coal gas used to light public streets in Pall Mall, London	1831—Englishman Michael Faraday invents the first electric generator	1807—Robert Fulton, credited with commercializing steamboat technology, builds his first steamboat, the Clermont, powered by wood. Wood was the fuel of choice for steamboats until the mid-1830s when coal came into use	1803—Louisiana Purchase
1830—First public railroad in the world driven by coal-powered steam locomotives, the Liverpool and Manchester Railway in England	1833—Small amounts of oil collected from seeps near the Allegheny River in Pennsylvania is bottled and sold as Seneca Oil medicine for fifty cents a bottle	1810—First U.S. patent for a gas lamp awarded to David Melville of Newport, R.I.	1831—American Joseph Henry invents the first electric motor		1804—Lewis & Clark Expedition
During the 1830s—Cast iron coal stoves began to replace open fireplaces for heating and cooking in American homes	1849—Canadian geologist Abraham Gesner distills kerosene from oil	1812—London and Westminster Gas Light and Coke Company chartered as the first coal gas utility in the world. Its purpose was lighting public streets in London	1834—Electric streetcar is invented by Thomas Davenport, blacksmith and inventor from Vermont	1828—Total consumption of fuels in the United States is approximately 80 percent wood, 3 percent charcoal, 2 percent bituminous coal and 14 percent anthracite coal	1814—Francis Scott Key writes the Star Spangled Banner after watching the British bombard Ft. McHenry in Baltimore
By 1831—The port of Pottstown, Pa., well established in shipping high-quality anthracite coal to Philadelphia, via man-drawn canal boats			1844—First electric telegraph built by American inventor and artist Samuel Morse	1830—First American steam locomotive in use on the Charleston & Hamburg Railroad in New York	1819—First section of Erie Canal is opened
					1830—Joseph Smith publishes The Book of Mormon
					1838—Trail of Tears. As a result of the controversial Indian Removal Act of 1830, 17,000 Cherokees were forced to emigrate from Georgia to Oklahoma. At least 4,000 perished in removal camps and on the trail

1801–1850 (continued)

Coal	Oil	Gas (Coal Gas and Natural Gas)	Electricity	Nonfossil Energy	Current Events
1835 to 1855—Pennsylvania's industrialization explodes with its capability to smelt iron ore directly from local supplies of low-sulfur anthracite coal		1816— Gas Light Company of Baltimore founded. It is the first energy utility in the United States, and Baltimore is the first U.S. city to use coal gas for street lighting. Gas lighting would come to Boston in 1822, New York City in 1825, New Orleans in 1833, and Philadelphia in 1836		1830— Baltimore & Ohio Railroad is the first to be established	1842—The New York Philharmonic, the first U.S. orchestra, is founded
1839—Invention of the steam shovel by American William Otis makes surface mining an efficient method for removal		1820s—Village of Fredonia, N.Y., captures seeping natural gas in a pipeline to use for lighting. The Fredonia Gas Light Company is the first natural gas company in the United States		By 1831—There were over 900 smelters of iron ore in the United States, all using charcoal	1846—United States declares war with Mexico
1840s—First industrial-scale coke ovens operating in United States				1834—McCormick patents mechanical reaper, which allows more efficient use of muscle power	1849—California Gold Rush begins when gold is discovered at Sutter's Mill
1846—Canadian Abraham Pineo Gesner discovers process for refining coal to make kerosene (also called coal oil)					

1851–1900

Coal	Oil	Gas (Coal Gas and Natural Gas)	Electricity	Nonfossil Energy	Current Events
1853—French chemist Charles Gerhardt derives acetylsalicylic acid from coal tar. Felix Hoffmann of the Bayer Company in Germany refines the formula and Bayer markets the drug in 1899 as aspirin. Salicylic acid used as a pain remedy was previously refined from willow bark	1852—Polish pharmacist Ignacy Lukasiewicz distills kerosene from oil and begins oil industry in Europe.	1850s—Coal gas cooking stoves begin to be manufactured but would not become popular until the 1890s	1851—Expansion cycle refrigeration system patented by American John Gorrie	1857—Pullman sleeping car invented	1860—Abraham Lincoln elected President
1856—First synthetic dye was the color mauve, developed from aniline from coal tar by English chemist William Perkin	1854—Pennsylvania Rock Oil Company founded, the first oil company in the United States	1854—Coal gas lighting arrives in San Francisco. Because of the lack of local coal, supplies were imported from Australia	1852—First commercially viable elevator developed by Elisha Otis	1869—Transcontinental Railroad completed with East meeting West at Promontory Point, Utah. The Central Pacific locomotive "Jupiter" (from the west) was burning wood and the Union Pacific locomotive #119 was burning coal.	1861—First Battle of Bull Run
By 1859—Almost 300 small coal gas plants were scattered across the country serving almost four million municipal, residential, and small industrial customers	1855—Connecticut Chemist Benjamin Stillman distills petroleum, producing gasoline, tar, and solvents	1854—A well drilled for water in Stockton, Calif., produces natural gas that is used to light the Court House	1879—Edison perfects the incandescent bulb	1888—The first large "wind turbine," built by George Brush, produces electricity in Cleveland, Ohio	1862—Abraham Lincoln signs the Pacific Railway Act, authorizing the first transcontinental railway
	1857—American Michael Dietz invents a lamp that can burn kerosene distilled from coal	1859—Natural gas discovered at Titusville, Pa., which was piped to Pittsburgh for industrial uses.	1882—Edison opens Pearl Street Station in New York City, the first coal-fired, central electric generation station, initially to power lighting	1890s—The first commercial solar water heaters were sold in California	1864–1866—Scottish physicist James Maxwell establishes modern physics with his work on the kinetics of gasses, and the connections between light, electricity, magnetism, and electromagnetic waves
	1859—U.S. Petroleum industry born in Titusville, Pa, with the first commercially successful oil well	1885—Robert Bunsen invented the Bunsen burner, which could produce a controllable flame, expanding the uses of gas beyond lighting	1882—First Hydroelectric Plant in operation on the Fox River, Appleton, Wisconsin. By 1889 there were 200 hydroelectric plants in the United States	1896—Henry Ford invents his first "horseless carriage," the quadricycle, which runs on pure ethanol	1865—Lee surrenders at Appomattox
			1883—Elevated electric commuter train begins operation in Chicago		1866—The first transatlantic telegraphic cable laid, making rapid communication with Europe possible

1851–1900 (continued)

Coal	Oil	Gas (Coal Gas and Natural Gas)	Electricity	Nonfossil Energy	Current Events
1866—First surface mining for coal occurs at Danville, Ill.	1870—John D. Rockefeller founds the Standard Oil Company	1886—Auer von Welsback, Austrian scientist, invents the gas mantle, a small cloth bag impregnated with rare earth elements that burns away leaving a fragile ash that glows as the gas is combusted	1884—British engineer Charles A. Parsons develops the steam turbine for electric generation		1867—The United States purchases Alaska from Russia
By the 1870s—Railroad locomotives were transitioning from wood to coal to fire their steam engines.	1885—German engineer Karl Friedrich Benz develops first gasoline-powered automobile engine.		1886—Schuyler Wheeler invented the electric fan, the only cooling device available for homes until air conditioning is invented		1867—The typewriter is invented
1873—The first cable cars went into operation in San Francisco. They were powered by coal-fired steam engines in the powerhouses. In 1901, the fuel was switched to oil, and in 1912, the steam engines were replaced by electric motors	1886—The first modern oil tanker is built in Germany	1887—Pennsylvanian Solomon R. Dresser patents a leak-proof gas line coupling, a design that remains almost unchanged today			1867—Alfred Nobel patents dynamite
	1889—Charter Gasoline Engine Company of Sterling, Illinois, manufactures first gasoline-powered farm tractor	1891—A 120-mile-long natural gas pipeline was built to transport gas from central Indiana to Chicago			1876—Custer defeated by Sioux at Little Big Horn
1890—United Mine Workers of America was founded					1876—Alexander Graham Bell patents the telephone
By 1890—The United States leads the world in coal production					1885—First skyscraper in Chicago
					1898—Spanish American War

1900–1920

Coal	Oil	Gas (Coal Gas and Natural Gas)	Electricity	Nonfossil Energy	Current Events
1902—Making history in labor relations, 150,000 anthracite coal miners go on strike causing a serious energy shortage and bringing in President Theodore Roosevelt, who helps settle the dispute in favor of the miners	1901—Spindletop Oil Field in Texas, the first big "gusher," increased the petroleum supply enough to advance automobiles as an economical form of travel	1910—Process discovered for manufacturing anhydrous ammonia from natural gas, the raw material for synthetic nitrogen fertilizers	1900—Municipally owned utilities provide about 8 percent of total electricity generated, used primarily for streetlights and trolley lines	1908—Henry Ford builds the Model T, a flex-fuel vehicle that can run on gasoline or ethanol	1900—Theodore Roosevelt is elected President
	1901—Oil drilling begins in Iran (then known as Persia). British get involved in 1907	1918—First major discovery of natural gas in the Texas Panhandle, followed by a similar discovery nearby in Kansas in 1922, formed the Panhandle/Hugoton Field, which would provide 16 percent of U.S. natural gas reserves	1902—First turbines in use to generate electricity	By 1910—Horse-drawn transportation is virtually gone from U.S. and European cities	1905—Albert Einstein comes up with $E=mc^2$, the Law of Mass-Energy Equivalence
1910—Bakelite, the first synthetic plastic, derived from coal tar (a waste product in the production of coke) invented by Belgian-American Dr. Leo Baekeland	1903—The Ford Motor Company is founded		1902—Willis Haviland Carrier patented the first air conditioner, used to stabilize temperature and humidity for industrial processes until the 1920s, when department stores began using them	1919—Ethanol, in use for decades as a lighting fuel, is banned during Prohibition unless it is mixed with petroleum	1906—Pure Food and Drug Act is passed
	1903—Frank and Orville Wright fly the first heavier-than-air craft at Kitty Hawk, North Carolina				1911—Mexican Revolution begins
1918—By the end of World War I, 75 percent of total energy use in United States is coal			1907—Georgia, Wisconsin, and New York establish public service commissions to maintain regulatory control over electric utilities		1913—17th Amendment to the U.S. Constitution is ratified, Federal income tax begins
					1914—The Panama Canal opens
					1917—U.S. enters World War I

1900–1920 (continued)

Coal	Oil	Gas (Coal Gas and Natural Gas)	Electricity	Nonfossil Energy	Current Events
1918—Pulverized coal is first used in elelctric generating plants; the method is still in general use today	1910—American chemist Walter Snelling is first to identify propane as a component of gasoline. It will eventually become a widely used byproduct of both petroleum and natural gas refining		1907—Portable electric vacuum cleaner invented and patented by James Murray Spangler, a Canton, Ohio, department store custodian. An early customer was a cousin, married to William H. Hoover. The rest is history		1920—The 19th Amendment to the U.S. Constitution passes, women get the vote
1920—Train travel peaks in United States at 95 percent of all intercity travel. This will drop 18% by 1929 because of rise of the automobile	1914—Standard Oil of California opens a chain of thirty-four gas stations		1907—Electric washing machine first mass-produced in the United States by the Hurley Machine Company of Chicago, Ill.		1920—First commercial radio broadcast
			1910—Neon light invented by French chemist Georges Claude		
			1913—The first geothermal power station begins operation in Italy		

1921–1930

Coal	Oil	Gas (Coal Gas and Natural Gas)	Electricity	Nonfossil Energy	Current Events
1923—U.S. employment in soft coal mining (bituminous and lignite) reaches all-time high at over 700,000	1921—The first drive-in restaurant is opened in Dallas, Texas	During the 1920s—Coal gas production peaks in the United States as electricity and natural gas became more widely available for lighting and industry	By the 1920s—20 percent of U.S. electricity is generated from hydropower	1920s–1950s—Solar water heaters are installed on over 50,000 homes in South Florida. Market disappears when cheap electricity and natural gas become available	1925—Scopes evolution trial in Tennessee
1923—Dragline excavators for surface mining are first used	1923—The Lincoln Highway is completed. It is the first coast-to-coast paved road, from New York to San Francisco	1923—Electric arc welding came into use for fabricating reliable long-distance natural gas pipelines	1927—The garbage disposal is invented by Wisconsin architect John W. Hammes		1927—Babe Ruth hits sixty home runs for the Yankees
	1925—The price of a Model T Ford drops to $290, equal to three months wages	1925—The first all-welded natural gas pipeline was built over 200 miles between Louisiana and Texas	1928—The Carrier Company offers the first residential air conditioner. The Depression and World War II slowed sales, but they picked up in the 1950s		1927—Al Jolson stars in *The Jazz Singer*, the first talking picture
	1926—First liquid fuel rocket is built and launched by R. H. Goddard, American physicist	1930—The first efficient gas water heater for home use was developed			1927—Charles Lindbergh flies solo across the Atlantic
	1928—The first diesel electric passenger locomotive is built. It could travel safely up to sixty-three miles per hour				1929—In October the stock market crashes marking the beginning of the Great Depression

1921–1930 *(continued)*

Coal	Oil	Gas *(Coal Gas and Natural Gas)*	Electricity	Nonfossil Energy	Current Events
	1930—First jet engine is patented by Englishman Frank Whittle 1930—Major oil fields discovered in East Texas 1920s–1940s—Development of new materials from petroleum byproducts include polyvinyl chloride (plumbing pipes), polystyrene (Styrofoam), and polyethylene (packaging materials). Also nylon, acrylics, and polyester				

1931–1940

Coal	Oil	Gas (Coal Gas and Natural Gas)	Electricity	Nonfossil Energy	Current Events
	1933—The first drive-in movie theater opens in Camden, N.J. By 1957 there would be 3,700 drive-ins in the United States	1935—The Public Utility Holding Company Act is passed to organize regulation of electric and gas utilities, breaking up multistate holding companies	By 1932—over 80 percent of urban homes have electricity while only 11 percent of farm homes do	1932—Physicist Ernest Orlando Lawrence, inventor of the cyclotron, discovers how to harness nuclear energy	1932—Franklin D. Roosevelt elected President
	1934—Streamlined, luxury comfort-level, diesel-powered trains revive interest in long-distance train travel. The Zephyr, featured at the 1934 Century of Progress Exhibition in Chicago, cut travel time in half compared to steam engines	1937—After an undetected natural gas leak caused an explosion in a Texas school that killed 294 people, Texas implemented gas odorization regulations that were later adopted nationally. Gas distributors now use mercaptan, an odorant smelling of rotten eggs	1933—Tennessee Valley Authority Act (TVA) establishes federal generating facilities to provide power to public-owned utilities and nonprofit cooperatives	1920–1935—Over 6.5 million windmills are erected in the United States, primarily on farms, to pump water, run saw mills, and generate electricity	1933—Prohibition repealed
	1935—Wallace Carothers of DuPont invents nylon, the first synthetic fabric		1935—President Franklin Roosevelt signed an executive order creating the Rural Electrification Administration	1939—Massachusetts Institute of Technology builds a solar-heated house	1933—Hitler becomes Chancellor of Germany
			1937—Bonneville Dam, the first federal hydroelectric dam, begins operation on Columbia River, and Bonneville Power Administration is established		1935—Social Security Act passed
					1935—First major league baseball game at night, Cincinnati, Ohio
					1936—Spanish Civil War begins
					1939—Germany Invades Poland

1931–1940 *(continued)*

Coal	Oil	Gas *(Coal Gas and Natural Gas)*	Electricity	Nonfossil Energy	Current Events
	1936—National City Lines, a holding company consisting of General Motors (GM), Firestone, and Standard Oil of California, begins buying out more than 100 electric trolley systems in 45 cities (including New York, San Francisco, Philadelphia, St. Louis, Salt Lake City, Tulsa, Baltimore, and Los Angeles) to be dismantled and replaced with GM buses. In 1949 GM and its partners were convicted of criminal conspiracy in this matter and fined $5,000		1939—The first digital electronic computer is designed by John Vincent Atanasoff and Clifford E. Berry		
	1938—Oil is discovered in Saudi Arabia and Kuwait				
	1940—Willys Motors introduces the Jeep, the first four-wheel drive vehicle for the Armed Forces				

1941-1950

Coal	Oil	Gas (Coal Gas and Natural Gas)	Electricity	Nonfossil Energy	Current Events
1942—Cyclone furnace developed, making it possible to burn lower grades of coal efficiently for electric generation and other steam processes	1940s—Most steam locomotives replaced by diesel and electric models in the United States	1940s—Natural gas replaces coal gas on Eastern Seaboard as large cities like Philadelphia, New York, and Boston convert their distribution systems	By the 1940s—40 percent of electricity comes from hydropower	1941–1945—Use of ethanol made from corn or other biomass sources increases during the war when petroleum supplies are tight	1941—Japan bombs Pearl Harbor and United States declares war
	1940s—Emergence of plastics industry and development of synthetics like nylon and polythene use oil as the raw material for production	1942—Natural gas liquefied for the first time, in Cincinnati, Ohio	By the 1940s—Trolleys have disappeared from American cities but retained widely in both Eastern and Western Europe. U.S. infrastructure for personal vehicles and bus lines is given municipal priority	1941—A one-megawatt wind turbine on Grandpa's Knob in Rutland, Vermont, produced power for the local community for eighteen months. The turbine was designed and built by a consortium of engineering companies and the local utility.	1944—Franklin Roosevelt is elected to his fourth term as President
	1946—The first drive-in banking windows open		1942—Manhattan Project begins to develop the atom bomb. This project leads directly to "peaceful" uses of nuclear technology for electric generation		1945—Atom bombs dropped on Hiroshima and Nagasaki
	1948—Farmer Frank Zybach, Colorado, invents the central pivot irrigation machine, a motorized irrigator that moves sprinkler arms in a circular pattern around a hub		1945—Microwave oven patented by American Perry L. Spencer		1946—Baby Boom begins
					1947—Jackie Robinson breaks the color barrier in major league baseball
					1947—First piloted supersonic flight, by Charles Yeager at Muroc Air Force Base, Calif.

1941-1950 *(continued)*

Coal	Oil	Gas *(Coal Gas and Natural Gas)*	Electricity	Nonfossil Energy	Current Events
	1948—The first Mc-Donald's Restaurant opens in San Bernardino, Calif.		By 1945—Close to half of U.S. farms have electricity		1948—Israel becomes a nation
	1949—Oldsmobile introduces the Rocket 88, the first V-8 automobile engine		1947—Transistor is invented by engineers at Bell Laboratories		1949—Mao leads Communist takeover in China
	1945–1950—Railroad passenger business plummets after the war, federal government support moves to roads and airports				1950—Korean War begins
					1950—United Nations Secretariat headquarters is built in New York

1951–1960

Coal	Oil	Gas (Coal Gas and Natural Gas)	Electricity	Nonfossil Energy	Current Events
1951—119 miners die as the result of an explosion in the Orient Number 2 mine in West Frankfort, Illinois. Congress passes the Coal Mine Safety Act of 1952, giving the Bureau of Mines authority to make inspections	By the mid-1950s—Railroads have moved from coal to diesel to power train engines	1954—First use of plastic pipe for transporting natural gas long distances. Materials were PVC, PE, and polybutylene	1951—The first UNIVAC Computer is delivered to the U.S. Census Bureau. This computer used vacuum tubes and stored data on magnetic tape	1953—President Eisenhower delivers the "Atoms for Peace" address to the United Nations, proposing an international Atomic Energy Agency to oversee world development of nuclear energy	1952—Dwight D. Eisenhower elected President
1952—Air inversion in London concentrates coal combustion smog, killing 4,000 people	1952—First jet air passenger service begins, running between London and Johannesburg	1956—The American Gas Association reports that natural gas has become the number one home heating fuel in the United States, providing service to 10.2 million residences	1951—The federal government's Experimental Breeder Reactor I in Idaho generates the first nuclear electricity	1954—First solar electric cell is developed at the Bell Telephone Laboratories	1954—Supreme Court rules against racial segregation in schools in *Brown vs. Board of Education*
	1956—The Interstate Highway Act creates the national highway network, ostensibly for national security	1959—Liquid natural gas (LNG) is produced in California at industrial scale for the first time. Gas is exported to England	1957—The first nuclear power plant in the United States to provide power to residential customers is in Shippingport, Pa.	1954—The Nautilus, the first nuclear-powered submarine is launched by the U.S. Navy	1955—Polio vaccine is developed by Dr. Jonas Salk and made available to the public
	1956—American Geophysicist Marion King Hubbert publishes his theory relating oil field production to a bell curve, and predicts U.S. oil production would reach its peak between 1965 and 1970 and would then continue to decline		1958–1959—Integrated circuit technology is developed by engineers at both Texas Instruments and Fairchild Semiconductor	1954—The Atomic Energy Act invites private industry into development of peaceful uses for nuclear power	1955—The Air Pollution Control Act, the first federal air quality legislation, is passed
					1955—Martin Luther King, Jr. begins the Montgomery Bus Boycott
					1956—The Soviet Union crushes the Hungarian revolt

1951–1960 *(continued)*

Coal	Oil	Gas *(Coal Gas and Natural Gas)*	Electricity	Nonfossil Energy	Current Events
	1960—The Organization of Petroleum Exporting Countries (OPEC) is founded in Baghdad. Countries include Iran, Iraq, Kuwait, Saudi Arabia, and Venezuela			1954—First transistor radio introduced by Texas Instruments	1957—The Soviet Union launches the Sputnik satellite
	1960—The first auto emissions law is passed in California, followed by the Motor Vehicle Air Pollution Act in 1965, which sets national standards			1955—The town of Arco, Idaho, becomes the first town to be completely powered by nuclear energy, from the Atomic Energy Commission's nearby testing station	1958—The first transatlantic jet passenger service is established by British Overseas Airways Corporation (BOAC), between New York and London
				1957—The International Atomic Energy Agency is founded to promote peaceful uses of nuclear energy	1959—Fidel Castro takes over Cuba
					1959—Alaska and Hawaii become states numbered forty-nine and fifty
					1960—John F. Kennedy is elected President

1961–1970

Coal	Oil	Gas (Coal Gas and Natural Gas)	Electricity	Nonfossil Energy	Current Events
1969—Federal Coal Mine Health and Safety Act is signed by President Nixon. This act amended the 1952 law setting stricter inspection standards and created a federal agency to oversee mine health and safety	1960s—Introduction of catalytic converters reduced hydrocarbon emissions of automobiles	During the 1960s—Market for manufactured coal gas disappears and remaining refineries close	1962—American Nick Holanyak, Jr. invents the light-emitting diode (LED)	1960s—Nuclear power industry blossoms, achieving 1% of U.S. generation by 1970	1961—Peace Corps is established
	1966—Electronic fuel injection invented, replacing the carburetor and increasing engine efficiency		1963—General Electric introduces the first self-cleaning oven		1961—Americans Alan Shepard and Gus Grissom are launched into space, after Soviet Cosmonaut Yuri Gagarin had made the trip as the first man in space
	1968—Use of synthetic fibers overtakes use of natural fibers in the United States. Most synthetics are produced from petroleum		1967—First hand-held calculator developed by Texas Instruments, weighs almost three pounds		1963—Civil Rights March on Washington, 250,000 in attendance
	1968—Oil is discovered on Alaska's North Slope		1968—Douglas Engelbart of the Stanford Research Institute demonstrates his invention, the computer mouse, at a computer show in San Francisco. He patents the device two years later		1963—President Kennedy assassinated
	1970—"Peak Oil." U.S. oil production peaks, as predicted by M. King Hubbert in 1956		1968—Wild and Scenic Rivers Act is signed, which designates rivers to be preserved from hydroelectric power projects		1963—U.S. Postal Service begins using zip codes
					1965—American combat troops are sent to Vietnam
					1966—Endangered Species Act signed
					1966—National Organization for Women (NOW) founded

1961–1970 *(continued)*

Coal	Oil	Gas *(Coal Gas and Natural Gas)*	Electricity	Nonfossil Energy	Current Events
			1969—The National Environmental Policy Act requires utilities to prepare environmental impact statements (EIS) as part of the process to obtain a permit for a new generating plant		1967—Arab-Israeli six-day war
					1967—Air Quality Act is signed into law
			1970—First CD-ROM developed by James T. Russell at Pacific Northwest Laboratories, Richland, Washington. Audio companies would acquire licenses in the mid-eighties		1968—Martin Luther King and Robert Kennedy are assassinated
					1968—National Democratic Convention in Chicago marked by anti-Viet Nam War protests and police violence
					1969—Apollo crew lands on the moon
					1969—Woodstock Music and Art Festival, 400,000 attend
					1970—Kent State Massacre at Kent State University, where the Ohio National Guard fired on student war protesters, killing four and wounding nine
					1970—Clean Air Act passed

1971–1980

Coal	Oil	Gas (Coal Gas and Natural Gas)	Electricity	Nonfossil Energy	Current Events
1972—Ninety-one miners perish from smoke inhalation and carbon monoxide fumes in the Sunshine Mine fire in Shoshone County, Idaho	1973—OAPEC declares the oil embargo that would dramatically change U.S. energy habits and policy	By 1970—Plastic pipe distribution lines in United States for natural gas total over 45,000 miles	1971—Intel invents the "computer chip" microprocessor, which is used in the Pioneer 10 spacecraft sent to survey Jupiter	1970s—Interest grows in ethanol as a transportation fuel or additive	1971—Federal government lowers voting age to eighteen
1980—National Acid Precipitation Assessment Program Study enacted to assess the impact of "acid rain"	1974—Automobile air bags become available as an option	1971—Natural gas productivity in the United States peaks after rapid market expansion between the 1940s and 1960s. Domestic production declines	1972—Magnavox markets Odyssey 100, the first home video game	1971—First offshore wind farm in operation off the coast of Denmark	1971—President Nixon visits China
	1975—The Energy Policy and Conservation Act is signed, which includes vehicle efficiency standards and establishment of a strategic oil reserve	1978—The Natural Gas Policy Act establishes the Federal Energy Regulatory Commission (FERC), which is given the authority to regulate the interstate gas industry	1975—The Altair 8800, the first home computer, is marketed by Micro Instrumentation Telemetry Systems	1972—Students at UCLA convert a 1972 Gremlin to run on hydrogen and win first prize for lowest tailpipe emissions in the 1972 Urban Vehicle Design Competition	1972—Break-in at Democratic headquarters at the Watergate Hotel eventually leads to investigation and indictments of members of President Nixon's administration
			1977—Apple Computer markets the Apple II computer, the first computer with keyboard, monitor, and mouse	1977—First human-powered aircraft, the Gossamer Condor, created by aeronautical engineer Paul MacCready	1973—Peace agreement signed in Paris between North and South Viet Nam, and United States leaves South Viet Nam
			1977—Major power blackout in New York City	1977—The Solar Energy Research Institute (SERI) is founded by the federal government to undertake research. It later becomes the National Renewable Energy Laboratory (NREL)	1973—Endangered Species Act is signed
					1974—President Nixon resigns after impeachment proceedings begin

Coal	Oil	Gas (Coal Gas and Natural Gas)	Electricity	Nonfossil Energy	Current Events
			1978—Singer markets the Athena 2000, the first sewing machine with electronic embroidery stitch program	1978—Last nuclear power plant ordered in the United States until the twenty-first century	1974—Freedom of Information Act is passed
			1978—National Energy Conservation and Policy Act signed. An important component was the Public Utility Regulatory Policies Act (PURPA), which required utilities to purchase power from qualified independent producers, a boon to small, renewable energy generators.	1979—Cooling unit fails at Three Mile Island nuclear plant in Harrisburg, Pa., causing a partial meltdown of the core and forcing 3,500 people to evacuate the area. The Three Mile Island nuclear plant accident led to sweeping changes in nuclear plant regulation	1976—Environmental Protection Agency bans pesticides containing mercury
					1977—President Carter signs an unconditional pardon of men who evaded the draft during the Viet Nam War
					1977—U.S. Department of Energy is established
				1979—Solar water heater sales balloon to 50,000 systems after the 1973 oil embargo	1978—President Carter declares Love Canal development in Niagara Falls, N.Y., a disaster area as the result of chemical pollution dumping
				1979—Federal government begins support for development of ethanol industry through subsidies and tax breaks	
				1980—First U.S. wind farm is built at Crotched Mountain, N.H., comprising twenty 30kW turbines	

1981–1990

Coal	Oil	Gas (Coal Gas and Natural Gas)	Electricity	Nonfossil Energy	Current Events
1983—The federal government commits its first funding, $2.5 billion, to subsidize development of clean coal technologies	1986—The deadline by which all lead was to be removed from gasoline. Ethanol was considered a potential substitute octane booster	1990—Use of natural gas increases in response to the Clean Air Act, primarily as a substitute for coal-fired electricity generation	1981—IBM PC introduced by Microsoft	1980s—Advancements in wind turbine design move toward large utility-scale machines and development of wind farms	1980—Mt. St. Helens erupts in Washington State, killing sixty-two and decimating hundreds of square miles of the landscape
1984—The Dakota Gasification Company's Great Plains Synfuels Plant near Beulah, North Dakota, goes on line to produce synthetic natural gas from coal	1989—Exxon Valdez runs aground in Alaskan waters, spilling eleven million gallons of crude oil into Prince William Sound		By 1982—Coal-fired power provides half the electricity in the United States. Hydropower, nuclear power, and natural gas each provided between 10 and 15 percent	1981—The Solar Challenger, the first solar-powered aircraft, is flown across the English Channel	1980—John Lennon is murdered outside his Manhattan apartment house
1984—The Willburg Mine fire claims twenty-seven lives in Emery County in Utah			By 1983—Solar electric generating stations in California are generating 13.8 megawatts of power for Southern California Edison	1983—President Reagan signs the Nuclear Waste Policy Act, the first legislation designed to deal with nuclear waste	1981—Sandra Day O'Connor becomes the first woman named to the U.S. Supreme Court
1986—Electronic gas detectors replace canaries in British coal mines			1984—Macintosh personal computer is introduced by Apple Computer	1984—First North American tidal power plant goes into operation at Annapolis, Nova Scotia, generating 20 megawatts of electricity	1981—Release of fifty-two American hostages, captured by Iran in 1980
1988—Wyoming replaces Kentucky as the state producing the most coal					1983—Sally Ride becomes first woman in space as part of Challenger space shuttle crew

1981–1990 (*continued*)

Coal	Oil	Gas (*Coal Gas and Natural Gas*)	Electricity	Nonfossil Energy	Current Events
			1987—The Montreal Protocol is opened for signature. It is an international agreement to eliminate use of CFC refrigerants, which have been found to be destroying the protective ozone layer	By 1985—California's wind turbines were producing over 1,000 megawatts of power. By 1990 that number would reach 2,200 megawatts	1984—Viet Nam Memorial in Washington, D.C., designed by architect Maya Lin, dedicated by President Reagan
				1986—Chernobyl nuclear power plant accident occurs. Explosions produce radioactive particulate clouds that move across Europe	1986—Challenger space shuttle explodes
					1987—Clean Water Act passed, overriding President Reagan's veto
					1988—Indian Gaming Regulatory Act allows gambling on reservations
					1989—Berlin Wall falls
					1989—President Bush, Sr. authorizes attack on Panama to seize President Manuel Noriega, accused of drug trafficking

Coal	Oil	Gas (Coal Gas and Natural Gas)	Electricity	Nonfossil Energy	Current Events
1995—The Wabash River Coal Gasification Repowering Project near West Terre Haute, Indiana, goes on line, This is one of the first IGCC Plants in the United States that uses coal gasification and combined cycle technology	1991—Japan moves into the number one spot in automobile production	1998—According to the U.S. Department of Energy, 5.1 billion cu ft of natural gas was being used for vehicles	1998—Printed plastic transistor process is developed by Bell Labs, making possible programmable cards and flexible computer screens	1992—The Energy Policy Act (EPACT) established a production tax credit for wind power at a rate of 1.5 cents per kilowatt hour. This credit would provide stabilizing support for the growth of the wind industry over the next fifteen years	1991—Public is given access to the World Wide Web
	1991—Operation Desert Storm is launched to drive Saddam Hussein from Kuwait and to protect the U.S. oil supply	2000—95 percent of new electricity generation in the United States is powered by natural gas	1999—Electricity is first marketed on the internet		1991—Clarence Thomas is confirmed to the Supreme Court after controversial confirmation hearings alleging sexual harassment
1996—Polk Power Station Unit #1 built and owned by Tampa Electric Company near Tampa, Florida goes on line using IGCC technology	1993—Oil imports exceed domestic production for the first time			1992—EPACT also defines "alternative transportation fuels" such as E-85, a fuel consisting of 85 percent ethanol	1992—The Energy Policy Act establishes minimum efficiency standards for commercial buildings
	1997—Toyota introduces the Prius, the first hybrid car to run on gasoline and electricity. Sales explode from about 18,000 in 1998 to almost 200,000 in 2005. The Prius would not be available in the United States until 2000			1995–2001—Solar pool heaters enjoy a strong market that grows 100 percent during this time.	1992—North American Free Trade Agreement (NAFTA) signed
					1992—Hurricane Andrew kills thirty-seven and causes $20 billion damage in Florida
	1997—The new Volkswagen Beetle model introduced			2001—660kW wind turbine is installed beside Hull, Mass., High School on Boston Harbor	1992—The Earth Summit convenes in Rio de Janeiro

1991–2000 (continued)

Coal	Oil	Gas (Coal Gas and Natural Gas)	Electricity	Nonfossil Energy	Current Events
	1997—Major U.S. automakers begin producing flex-fueled vehicles that can run on gasoline and E-85, a mixture 85 percent ethanol				1993—First terrorist attempt to destroy the World Trade Center in New York is a bombing that kills nine people and injures almost three hundred others
	1998—Chrysler Corporation purchased by Mercedes-Benz				1994—Nelson Mandela is elected President of South Africa
	1999—The Honda Insight becomes the first hybrid car available in the United States				1995—Federal building in Oklahoma City is bombed, killing 168
					1995—Microsoft's Windows 95 software is released
					1997—Astronauts recalibrate the Hubble space telescope in a 33-hour space walk

2001–2007

Coal	Oil	Gas (Coal Gas and Natural Gas)	Electricity	Nonfossil Energy	Current Events
2003—George Bush announces that DOE will begin development of FutureGen, the world's first integrated sequestration and hydrogen production power plant designed to be emission-free. The price tag is $1 billion	2004—The Chevy Silverado and the Dodge Ram, the first hybrid full-size pickup trucks become available, along with the Ford Escape, the first hybrid SUV	2001—Natural gas prices rise 50 percent	2001—An electricity shortage causes rolling blackouts in California and an energy crisis for Governor Gray Davis	2003—Total wind turbine capacity in the United States exceeds 6,300 megawatts, located primarily in California and Texas	2001—World Trade Center in New York is destroyed by suicide terrorists
2005—Federal Energy Policy Act provides major new resources for clean coal technology development	2005—Oil prices spike worldwide after Hurricane Katrina flattens New Orleans	2002—In the first non-stop balloon flight around the world, American Steve Fossett travels from Northland, West Australia, to Yamma Yamma, Queensland. The balloon is filled with helium and hot air (helium is derived from natural gas)	2003—The most extensive blackout in North American history occurs in mid-August in the northeastern United States and parts of Ontario, Canada, affecting 50 million people	2005—There are more than four million flex-fuel vehicles in the U.S. and over 400 E-85 filling stations, mostly in the Midwest	2001—Enron goes bankrupt
2006—Thirteen miners are trapped after an explosion at the Sago Mine in Tallmansville, West Virginia. After forty hours of effort, one survivor is rescued		2005—Hurricanes Rita and Katrina shut down natural gas production in the Gulf of Mexico, the source of close to one quarter of all U.S. production			2002—The Euro becomes legal tender for the dozen member nations of the European Union
		2006—A record of 31,687 natural gas wells are drilled			2002—Former President Jimmy Carter is awarded the Nobel Peace Prize
					2003—Seven astronauts die as space shuttle Columbia disintegrates when it reenters the earth's atmosphere
					2003—U.S. invades Iraq

2001–2007 (*continued*)

Coal	Oil	Gas (*Coal Gas and Natural Gas*)	Electricity	Nonfossil Energy	Current Events
					2004—Ten nations join the European Union including eight that were formerly under Communist rule
					2005—Lance Armstrong wins his seventh consecutive Tour de France and then retires from racing
					2005—Pope John Paul II dies at age eighty-four

Glossary of Electricity Terms

This Appendix offers explanation and definition of some terms in common use in the electricity industry, including units of measurement such as kilowatts and ohms, and utility management terms like load shaping and net metering. Common acronyms are also noted.

An Outline of Electricity

The electric energy delivered from the generating plant to the wall outlet is a type of wave energy like X-rays, microwaves, and telephone signals, only it vibrates at a different frequency. It is activated at the molecular level when normally neutral atoms are pulled apart by the presence of heat, light, or magnetism and then seek to regain their neutral state. This electromagnetic vibration creates electrical energy. The characteristics of the electrical energy generated for our use, which we call electricity, are measured in three different units called volts, amps, and watts.

Volts (voltage)

Voltage is the measurement of the intensity in the electrical field created between two points. A difficult concept to define, volts could be imagined as the units of electrical "oomph" between two points in a circuit. In the context of the electric power industry, voltage is used to rate the operating capacity of transmission or distribution lines and equipment.

Amps (ampheres)

Amps measure the amount of electric current, or the flow of the electric charge. A current of one amp is produced with one volt, steadily applied in a circuit with one ohm of resistance. Most homes are now wired for a 100-amp service, which means appliances and lights can draw up to 100 amps of current simultaneously before the circuit breaker trips.

Resistance (ohms)

Resistance, measured in ohms, is the degree to which a material resists the passage of electric current, thereby converting it to another form of energy, usually heat. It is resistance in the light bulb filament that causes it to heat up to the point of producing white-hot light. Much of the energy required to power a light bulb is lost as heat. Resistance in power lines is what causes losses of power to occur over distances.

Watts (wattage)

Watts are units of measurement for electric power or capacity. Watts combine the properties of volts and amps (watts = volts × amps) to produce a standard unit for measurement. Watts are shown in multiples of 1,000, as kilowatts (kW), megawatts (MW), and gigawatts (GW) (see table below).Watts can be a measurement of power production, as in the capacity of a generating plant (a 500 kW wind turbine), the transport of power along transmission lines (a line capable of moving 540 MW of energy per hour), or how much a specific electrical device uses (a 100 watt light bulb).

Watt (W) = 1
Kilowatt (kW) = 1,000 watts (1 × 1,000)
Megawatt (MW) = 1,000,000 watts (1,000 × 1,000)
Gigawatt (GW) = 1,000,000,000 watts (1,000,000 × 1,000)

Watt hour (Wh)

Watt hours measure electric energy, which is the commodity we purchase for use. One watt hour of electric energy will run a one-watt device for one hour. That is a pretty small amount, so watt hours are usually expressed in 1,000s, or kilowatt hours (kWh). One kilowatt hour of electric energy will run a 100 watt light bulb for ten hours. Utility customer rates are usually based on the price per kilowatt hour, and this is what appears on the customer's monthly bill. Production of electric energy from generation plants is usually expressed as megawatt hours or gigawatt hours.

Direct Current and Alternating Current (DC and AC)

Simply stated, direct current (DC) moves in one direction all the time, and alternating current (AC) reverses direction in a regular pattern or sine wave. Both types of current can be used, but AC became the U.S. standard in the early twentieth century because the technology was more efficient. AC produces a hum, which is the 60-cycle frequency of the sine wave, and which can be heard in fluorescent tubes. Solar electric panels generate DC power, which must be changed to AC by an inverter before it can be used in a conventionally wired home.

Other Electricity Terms

Base Load Plant—Base load plants provide the ongoing supply of power used every day, regardless of time or season. Slow to start up, these plants run continuously except for scheduled maintenance. Most modern base load plants are coal or nuclear-powered steam turbine generation plants ranging in size from one megawatt to over one thousand megawatts. Large hydropower plants (powered by the flow of a river) also provide base load power.

Combined Cycle Generator—A combined cycle generator increases fuel efficiency by harnessing the waste heat in the exhaust gasses from gas combustion turbines. However, instead of capturing the heat for use outside the generator, it is redirected internally to a boiler that powers a conventional steam turbine to generate additional electricity.

Combined Heat and Power (CHP) Systems—Also known as cogeneration, CHP increases the efficiency of the fuel being used to generate electricity by harnessing the heat produced by the generator. This means capturing the heat for use in an industrial process or for space heating. Because CHP is most efficient when all the heat can be used, systems are usually designed to satisfy the heating requirements rather than the electrical load.

Demand Side Management (DSM)—Utilities must meet the demand for electricity as it happens so they employ strategies to exercise some advance control over when and how much demand there will be. This is Demand Side Management, which, in combination with Supply Side Management (SSM) (managing bulk power purchases or acquiring new generation capacity), comprises "load shaping." The less variation in the overall load, the more economical it is to supply the power for it. Demand side management strategies include conservation programs designed to reduce energy use during particular times of the day or season, time-of-use rates to encourage less use during peak hours, and interruptible load rates, usually offered to large industrial energy users who must then shut down on request during peak load times or emergency power shortages. Utilities can also promote technologies that make use of energy generated off peak, such as ice storage systems that make ice at night when rates are low, then use it in chillers during the daytime when rates are high.

Electric System Losses—Much of the fuel energy that enters the generation process gets transformed into heat rather than electricity. In a coal-fired power plant, up to 70 percent of the energy in the coal is lost by the time the electric power leaves the plant, and another 8 percent is lost along transmission and distribution lines. The clouds of steam emerging from the stacks of a coal-fired plant represent heat energy from the coal being burned that isn't being used to make electricity, but is simply released to the air. It is referred to as "waste heat."

Generator—A generator converts mechanical energy to electrical energy, usually through use of magnetic energy. A magnet is moved across a coil of wire, creating an electromagnetic vibration that generates an electric current. The mechanical energy source typically drives a shaft that rotates the magnet inside the coil of wire. This source could be an internal combustion engine or turbine. Most generators today are turbine driven. Turbines are powered directly by water or wind, or by steam produced in coal-, diesel-, or nuclear-powered boilers, or by hot gasses produced by burning natural gas under pressure.

Independent Power Producer (IPP)—An entity, private or public, that generates electricity for sale on the wholesale power market but is not classified as a utility. An example would be a wind farm development firm that builds and then operates large-scale wind farms to produce bulk power for sale.

Interconnection Agreement—When the owner of a generation unit or system wants to connect to the grid, he or she must sign an interconnection agreement with the utility that owns the line. The agreement usually includes clauses regarding schedule and rates paid for the electricity, safety guarantees for utility line employees, insurance requirements, location of a disconnect switch, and other responsibilities of both parties.

Internal Combustion Generator—Internal combustion generators were used before the invention of the much simpler and more efficient turbine generator. Burning fuel drives the cylinders that operate the shaft connected to the generator. There are practical limits to the size of this technology, which is why their use is limited to smaller applications like portable generation units and small peaking plants. Typically powered by diesel, internal combustion generators are also used with anaerobic digesters to generate electricity from methane in sewage or animal waste.

Load Shaping—Utilities attempt to match ongoing electric supply with electric demand through load shaping strategies. The term comes from the goal of making the generally consistent shape of electric supply conform with the more unpredictable shape of electric demand. These strategies fall into the two categories of "supply side" and "demand side" management. Supply side strategies involve the management of purchasing or generation of base load and peaking power to meet projected needs. Demand side management engages the customer in the load shaping process by offering rate incentives to shift or reduce the customer's load during peak demand periods.

Net Metering—In order to encourage the interconnection of renewable energy generation systems, net metering laws have been enacted in most states. With net metering, surplus electricity that is generated by the interconnected system is credited at the same rate the customer is paying for electricity from the utility. So, when the system is generating more than needed, the meter essentially runs backward, and when the system is not generating enough, the customer is drawing electricity from the utility grid and the meter moves forward. The system size limits vary considerably from state to state, but they are usually designed to accommodate small-scale, non-utility systems.

Peaking Plant—Peaking (or peak load) plants can be quickly brought on line during times of peak demand, those times of day or seasonal occurrences of greater electricity use, such as hot summer afternoons when the air conditioning load increases and demand for electricity exceeds base load plant production. Peak load power is more expensive than base load power because it costs more to produce. Typical peaking plants are older, less efficient coal- or oil-fired plants, or natural gas–powered plants built specifically for the purpose.

Renewable Energy Portfolio Standards (RPS)—A growing number of states are enacting statutes that require retail utilities to have a certain percentage of renewable energy generation in their supply portfolios by a certain date. Utilities can either own the generation or can purchase renewable generated power from other utilities. Some states require the portfolio to be broken down in specific percentages by technology as well.

Tariffs—Tariffs are the published electricity rates for different sectors and classes of electricity customers as determined by the regulators, and the general terms and conditions under which they apply.

Turbine Generator—Most electric generation today employs some type of turbine. A turbine consists of blades mounted on a shaft. The blades are moved by a fluid or gas, rotating the shaft that operates the generator. Coal and nuclear plants or plants that burn wood or municipal solid waste produce steam to drive the turbines. Natural gas–powered turbines are turned directly by the hot gasses produced in the combustion process. Hydroelectric turbines are turned either by the power of falling water or by the river current. Wind turbines are turned by the wind.

INDEX

Database of Incentives for Renewables and
Efficiency (DSIRE), 212
daylighting, 150–151
Demand Side Management (DSM). *See*
energy efficiency
diesel, 23, 263
distributed generation, 95–96; and energy
security, 203
district heating and cooling, 125, 171, 182,
192, 199–200; and heat pumps, 173
Dresser, Solomon (gas line coupling), 57
dynamo, 77, 96

Earthships, 119, 130
Earthwood Building School, 130
Edison, Thomas, 77–79
Edison Company for Isolated Lighting, 78
Edison Electric Institute (EEI), 98
Eisenhower, President Dwight, 63
electric lighting: compact fluorescent
(CFL), 101; incandescent bulbs, 77, 101;
Light Emitting Diodes (LEDs), 188;
street lighting, 76
electric motors, 77
electric power generation, 81; base load, 77,
78, 81, 262, 263; combined cycle, 261;
demand for, 82, 94; efficiency, 108; generator, 263; independent power producer
(IPPs), 81, 263; Integrated Resource
Planning (IRP), 112, 113, 114; on-site, 79,
95; peak load, 262, 264; peaking plant,
81, 264; pollutant capture, 37, 94; reliability of, 80, 87; system losses, 262–263
electric power system, centralized, 77, 79,
84, 93, 94, 95; Pearl Street Station, 78, 108
electric utility industry, 26, 77, 82–88, 92, 95;
community relations, 114, 209; future
trends, 97; industrial rates, 114; load
management, 114, 165, 262, 263–264;
"natural monopoly," 88, 90, 97, 112;
service territories, 88, 111; Supply Side
Management (SSM), 113, 262, 263; time
of use rates, 262; unbundling, 84, 92;
utility planning, 82, 209. *See also* utilities
electricity, 73–98; alternating current (AC),
78, 139; amps (amperes), 260; direct
current (DC); 78, 138, 261–262; glossary

of terms, 260–264; ohms (resistance),
261; volts (voltage), 260; watt hour, 261;
watts (wattage), 261
electricity deregulation, 90–93, 95, 97, 112,
114; community aggregation, 97
electricity distribution, 82, 93, 203; microgrids, 96
electricity regulation, 88–90; by states, 80,
84, 88, 91, 92, 111, 112; federal, 89; green
energy, 91, 95, 96, 187; interconnection,
89, 90, 91, 96, 263; local, state and federal
roles in, 89–90; net metering, 91, 96, 264;
renewable energy portfolio standards
(RPS), 91, 96, 140, 209, 215, 217, 264; tariffs, 91, 264
electricity transmission, 82, 89, 92, 203;
losses, 263; wheeling, 92
energy codes, 115–118; ASHRAE 90.1, 116–117,
198; International Energy Conservation
Code (IECC), 117–118, 195; Model Energy
Code, 117; state adoption of, 117. *See also*
energy efficiency
energy conservation, 100, 126–129
Energy Conservation and Production Act,
117
energy efficiency, 95, 100–129, 132; appliance
standards, 108–110, 118, 223; building audits, 107, 112; building codes, 101, 115–118;
Demand Side Management (DSM), 101,
112, 113–114, 194, 262, 263; federal policy
regarding, 101, 108, 215; in commercial
buildings, 198; in downtown revitalization, 199–200; residential, 113; standards,
102; utility programs, 112
Energy Efficiency and Renewable Energy
(EERE). *See* U.S. Department of Energy
energy efficient mortgages, 197
energy independence. *See* energy self-
reliance
Energy Independence and Security Act of
2007, 215, 224
energy infrastructure, 7, 133, 134
energy policy, 213–226; federal level, 214–215;
international collaboration, 220–221,
223; local government level, 215, 218–220;
regional efforts, 217–218; state government level, 215–218